Advances in Spatial Science

Springer
Berlin
Heidelberg
New York
Hong Kong
London
Milan
Paris
Tokyo

Titles in the Series

C. S. Bertuglia, M. M. Fischer and *G. Preto* (Eds.)
Technological Change, Economic Development and Space
XVI, 354 pages. 1995. ISBN 3-540-59288-1
(out of print)

H. Coccossis and *P. Nijkamp* (Eds.)
Overcoming Isolation
VII, 272 pages. 1995. ISBN 3-540-59423-X

L. Anselin and *R. J. G. M. Florax* (Eds.)
New Directions in Spatial Econometrics
XVX, 420 pages. 1995. ISBN 3-540-60020-5
(out of print)

H. Eskelinen and *F. Snickars* (Eds.)
Competitive European Peripheries
VIII, 271 pages. 1995. ISBN 3-540-60211-9

J. C. J. M. van den Bergh, P. Nijkamp and *P. Rietveld* (Eds.)
Recent Advances in Spatial Equilibrium Modelling
VIII, 392 pages. 1996. ISBN 3-540-60708-0

P. Nijkamp, G. Pepping and *D. Banister* (Eds.)
Telematics and Transport Behaviour
XII, 227 pages. 1996. ISBN 3-540-60919-9

D. F. Batten and *C. Karlsson* (Eds.)
Infrastructure and the Complexity of Economic Development
VIII, 298 pages. 1996. ISBN 3-540-61333-1

Y. Leung
Intelligent Spatial Decision Support Systems
XV, 470 pages. 1997. ISBN 3-540-62518-6

C. S. Bertuglia, S. Lombardo and *P. Nijkamp* (Eds.)
Innovative Behaviour in Space and Time
X, 437 pages. 1997. ISBN 3-540-62542-9

A. Nagurney and *S. Siokos*
Financial Networks
XVI, 492 pages. 1997. ISBN 3-540-63116-X

M. M. Fischer and *A. Getis* (Eds.)
Recent Developments in Spatial Analysis
X, 434 pages. 1997. ISBN 3-540-63180-1

R. H. M. Emmerink
Information and Pricing in Road Transportation
XVI, 294 pages. 1998. ISBN 3-540-64088-6

P. Rietveld and *F. Bruinsma*
Is Transport Infrastructure Effective?
XVI, 384 pages. 1998. ISBN 3-540-64542-X

P. McCann
The Economics of Industrial Location
XII, 228 pages. 1998. ISBN 3-540-64586-1

L. Lundqvist, L.-G. Mattsson and *T. J. Kim* (Eds.)
Network Infrastructure and the Urban Environment
IX, 414 pages. 1998. ISBN 3-540-64585-3

R. Capello, P. Nijkamp and *G. Pepping* (Eds.)
Sustainable Cities and Energy Policies
XI, 282 pages. 1999. ISBN 3-540-64805-4

M. M. Fischer and *P. Nijkamp* (Eds.)
Spatial Dynamics of European integration
XII, 367 pages. 1999. ISBN 3-540-65817-3

M. M. Fischer, L. Suarez-Villa and *M. Steiner* (Eds.)
Innovation, Networks and Localities
XI, 336 pages. 1999. ISBN 3-540-65853-X

J. Stillwell, S. Geertman and *S. Openshaw* (Eds.)
Geographical Information and Planning
X, 454 pages. 1999. ISBN 3-540-65902-1

G. J. D. Hewings, M. Sonis, M. Madden and *Y. Kimura* (Eds.)
Understanding and Interpreting Economic Structure
X, 365 pages. 1999. ISBN 3-540-66045-3

A. Reggiani (Ed.)
Spatial Economic Science
XII, 457 pages. 2000. ISBN 3-540-67493-4

D. G. Janelle and *D. C. Hodge* (Eds.)
Information, Place, and Cyberspace
XII, 381 pages. 2000. ISBN 3-540-67492-6

P. W. J. Batey and *P. Friedrich* (Eds.)
Regional Competition
VIII, 290 pages. 2000. ISBN 3-540-67548-5

B. Johansson, Ch. Karlsson and *R. R. Stough* (Eds.)
Theories of Endogenous Regional Growth
IX, 428 pages. 2001. ISBN 3-540-67988-X

G. Clarke and *M. Madden* (Eds.)
Regional Science in Business
VIII, 363 pages. 2001. ISBN 3-540-41780-X

M. M. Fischer and *Y. Leung* (Eds.)
GeoComputational Modelling
XII, 279 pages. 2001. ISBN 3-540-41968-3

For information about further volumes please see end of the book

John R. Roy

Spatial Interaction Modelling

A Regional Science Context

With 7 Figures
and 1 Table

Springer

Dr. John R. Roy
Environmental, Transport and Urban Development Studies (ETUDES)
PO Box 96, Mallacoota
Victoria, 3892
Australia
johnmall@bigpond.com

ISBN 3-540-20528-4 Springer-Verlag Berlin Heidelberg New York

Cataloging-in-Publication Data applied for

A catalog record for this book is available from the Library of Congress.

Bibliographic information published by Die Deutsche Bibliothek
Die Deutsche Bibliothek lists this publication in the Deutsche Nationalbibliografie;
detailed bibliographic data available in the internet at *http.//dnb.ddb.de*

Springer-Verlag Berlin Heidelberg New York
is a part of Springer Science+Business Media
springeronline.com

Printed in Germany

Cover design: Erich Kirchner, Heidelberg

SPIN 10970911 42/3130 – 5 4 3 2 1 0 – Printed on acid-free paper

Preface

Spatial interaction modelling has a long tradition, especially in geography, where it was introduced as 'gravity' modelling in the early part of last century. The key purpose of such models is to determine the influence of 'spatial separation' between origins and destinations on either locational or travel choices of firms and individuals. By the 60's, statistical concepts, also used in physics, had been introduced to give the models a stronger methodological basis. However, it was not until the early 90's that these models were put into a more formal probabilistic framework, being potentially opened up as most-probable-state models to be used in conjunction with any theory underlying the analysis at hand.

Within this overall development, my particular niche, explored fully in this book, is to relate spatial interaction modelling to regional science, especially to spatial economics. In this continuing endeavour, there are four colleagues to whom I owe the most. The first is Paul Lesse, a former colleague from the then Division of Building Research, Commonwealth Scientific and Research Organisation (CSIRO), Australia. His strong mathematics/physics background allowed us to understand entropy within a very generalised framework, especially in terms of relevant transformations of the conventional entropy objective functions. As the 'actors' being modelled in spatial interaction models are in many cases 'economic' actors, workers in this field are inevitably confronted with acquiring a knowledge of spatial economics, even if, as with myself, this is not one's original discipline. For this daunting task, I needed one or two key colleagues with not only the virtue of patience, but possessing flexible and open minds, the key quality which regional scientists strive for. I have been very fortunate to have collaborated with two key guys who need no further introduction here, Börje Johansson and Geoffrey Hewings. With Börje, I was gently led into coming to terms with abstruse concepts such as 'utility', supply and demand, existence of equilibria and, finally, imperfect competition. Fortified by many glasses of Australian red wine, which Geoff had miraculously unearthed in the Urbana of the early 80's, I was introduced to the concept of Regional Science, as well as to the manifold variations and subtleties in regional input-output analysis, which have fascinated me ever since. If there is any colleague whose collaboration would have made this book much more rigorous, it must be Tony Smith, originally from the Regional Science Faculty, University of Pennsylvania. Tony's work has been a continual fount of inspiration to me, stimulating the development of key concepts of the unifying regional science framework for spatial interaction modelling, which is the book's key focus.

I was fortunate that John Brotchie, head of our former group at CSIRO, which also included Paul Lesse and David Batten, was vitally interested in extending the understanding of entropy models and their links with models based on utility concepts. John's successor, Peter Newton, continued support for this work. Nariida Smith, the leader of the CSIRO planning project within the group, encourages this modelling work wherever possible. I would also like to thank Hideo Nakamura and Yoshitsugu Hayashi from Japan, who invited me to spend a year at University of Tokyo in 1994-95, where I was able to meet many innovative regional scientists, including Takeo Ihara, Kiyoshi Kobayashi, Kazu Miyamoto, Hisa Morisugi and Takayuki Ueda, to name but a few. The broad multi-disciplinary focus of my Japanese colleagues, many of whom, like me, had civil engineering as their original field, was a great encouragement. Collaboration with Francisco Martinez, University of Chile, has been both inspiring and productive, yielding most of the content of the housing chapter in this book. There are many other colleagues whose work has influenced me strongly. Among these, I would like to highlight Åke Andersson, Martin Beckmann, Dave Boyce, Stewart Fotheringham, Masahisa Fujita, Britton Harris, Folke Snickars, Jean-Claude Thill, Huw Williams, Alan Wilson, and the late Giorgio Leonardi. All these scholars have been a source of inspiration and a cause for thanks - perceived limitations in my own work should not be ascribed to any of them! Regional science requires *par excellence* a multi-disciplinary perspective, in a context where most of us inevitably have one discipline further developed than others. This fact implies that open collaboration, freed from the often unspoken or unrecognised assumptions of a particular discipline, can be especially productive. *Vive la difference!*

Mallacoota, Australia, in September 2003

John R. Roy

Contents

Prologue

Spatial Interaction Modelling. Let us imagine a particular worker residing in a particular location in a city and a particular job existing in a different location in the same city. Then, 'spatial interaction' may be defined here as the following hierarchical choice process. The first step is the negotiation of a bilateral employment contract between the worker identified by his or her place of residence and the employer identified by his or her place of business. The second step is the daily process of commuting by the contracted worker to and from his or her place of work. The recognition of these two steps helps us to understand the dynamics of spatial interaction. In practice, the totals for the first step, *location decisions*, enable the usage of the available *building stock* for housing and employment to be assessed spatially, whereas the totals for the second step, *travel decisions*, have implications for usage of the *transport network* and its design.

The negotiation of a work contract is generally a much more *durable* decision than the choice of travel mode or route for the journey to and from work. Route choice, in particular, may change from one day to the next, depending on perceived traffic conditions or on the need to divert from the most direct route for shopping or personal service stops. At the same time, when a worker negotiates an employment contract he or she would normally include some *expectation* of commuting time within the overall decision on whether or not to accept the job. So, although the direction of the hierarchy is intuitively obvious (one cannot commute until one has obtained a job), feedback effects should be considered. For instance, the worker may have seriously underestimated the required commuting time between home and work, and may in the longer term consider changing job. Alternatively, some discrete change in the transport system, such as the opening of a major section of ring road, may suddenly make other work locations relatively more attractive. In a wider context, the worker may decide to change residence instead. With spatial interaction defined as a two-level hierarchical choice process with feedback, changes over time in the spatial distribution of activities (eg. households, jobs and facilities) and in the corresponding patterns of flows on the networks connecting these activities may be understood as a two-level *nested dynamic* process.

Under the above definition, changes in spatial interaction patterns may be classified in the field of *microeconomics* as *short run* changes. In other words, spatial interaction has been depicted as spatial adjustments within given *capacities* of the housing stock, the stock of jobs and the transport infrastructure. On the other hand, the creation of new capacity is a *long run* process. Although such processes have not usually been included under spatial interaction modelling, they will be mentioned briefly in this book. After all, capacity changes are intrinsically spatial phenomena, directly generating changes in the patterns of spatial interaction (location and corresponding travel) within the new capacities. However, as such capacity changes are quite slow processes, they normally generate states of dynamic disequilibrium, with changes in faster adjusting quantities, such as prices, periodically disturbing the approach towards long run equilibrium. A discussion of recent progress in the modelling of such states would alone require several books. Thus, this book merely outlines some key principles to be applied in understanding and modelling long run change.

Having defined spatial interaction in a rather novel way, we are challenged to develop an appropriate *modelling* framework for this process. Spatial interaction modelling arose from the field of *geography*, as well as being influenced by several concepts already used in physics. In general, spatial interaction models try to *project* the behaviour of a spatial system at a future point of time, upon the introduction of exogenous projected changes in the spatial distribution of activity capacities and in the properties and locations of the transport (and electronic) networks which connect these activities. The models are *probabilistic* rather than deterministic, with behaviour being constrained to replicate key observations instead of being based on idealised single criterion optimization. In fact, the most fundamental *behavioural constraint to observations* in the *probabilistic* model corresponds to the *objective function* which is being maximized (or minimized) in the analogous *deterministic* model. Also, with sufficiently large populations, the most probably macrostate is demonstrated to be *overwhelmingly* the most probable, approaching the *expected* result.

We can demonstrate asymptotic convergence, as the level of 'uncertainty' decreases, from the results of the probabilistic model towards those of the deterministic model. For the *travel decision* in the job choice/commuting example described above, the objective of the deterministic model is to obtain the flow pattern which minimizes the total travel cost (or time) between all origins and destinations. But, for the probabilistic model, the *observed* average travel cost in the system is applied as the right-hand value of the key behavioural constraint which the overall flow pattern must satisfy. Convergence occurs towards the flow pattern of the deterministic solution as the observed average travel cost approaches the cost-minimizing result. As commuters, in general, have more objectives in their choice of job than the minimizing of their commuting costs, a probabilistic approach seems more realistic in practice.

One of the current key debates in spatial modelling relates to the use of ***aggregate*** vs ***disaggregate*** models, where spatial interaction models are often dismissively assigned to the *aggregate* class. However, we argue here that this debate is usually based on false premises. Firstly, we note that the so-called aggregate models usually divide the analysis into market segments, rather than treating the total population in a single analysis. Note here, that as the number of such segments approaches the number of individuals in our disaggregate sample, the results of the 'aggregate' model become *identical* to those of the disaggregate model. So long as the market segmentation is cleverly performed such as to minimise the variance of the behaviour *within* each segment, the results of the two approaches can be quite close, even for a modest number of segments. Secondly, some data used in disaggregate models must be extracted from an aggregate analysis. For example, as traffic congestion is an aggregate phenomenon, travel times for individuals in peak periods must usually be obtained from an aggregate traffic assignment analysis! Also, because of confidentiality constraints, much of the data on the behaviour of firms is simply unavailable at the level of the individual firm. In addition, the process of aggregating up the results of disaggregate analyses for policy purposes is fraught with difficulties. Finally, although disaggregate demand models have proved their worth for problems with just one or two linked events (eg. joint choice of transport mode and destination), issues of appropriate sample size can become critical for multi-event problems. As shown in Chap. 4, this applies in particular to models of retail/service activity, where consumers make alternative multi-stop tours jointly with the choice of multiple goods and services at each stop. In order to ensure that the number of individuals in each entry of the multi-dimensional choice matrix is statistically not too small, a disaggregate approach must often partition the choice dimensions into several sequential subsets, thereby sacrificing the possibility of identification of interdependencies between the subsets. On the other hand, the more aggregate market-segmented models can often retain the full set of choices in the one analysis, without the number of entries in each element of the array becoming too small. At the same time, we note that it would be possible, in theory, to analogously define probabilistic *entropy* forms of our spatial interaction models at the level of the *individual* (Anas, 1983). This yields a structure similar to the well-known multinomial or nested logit models (Roy, 1985). For the reasons stated above, we have deliberately chosen not to follow this path. In this 'art' of modelling, we are inevitably confronted with compromises - 'black and white' characterisations based on extremes in model types help nobody!!!!!! In addition, we face the perennial challenge of choosing which variables are endogenous and which can be reasonably set as exogenous outside the model. Judgment based on experience is our main guide here.

A Regional Science Context for Spatial Interaction Modelling. Spatial interaction modelling, having arisen from the field of geography with borrowings from physics, uses insights from the two disciplines. Rather than narrowly interpreting this sharing as implying that 'people behave like molecules', we prefer to state that a common mathematical framework reliably describes group behaviour when close to equilibrium in both fields. 'Regional Science' is even more pluralistic, harnessing fresh insights from any discipline with something interesting to contribute to the understanding and modelling of spatial phenomena.

The most obvious omission in spatial interaction modelling is the discipline of economics, especially *microeconomics*. However, several researchers have already begun to include insights from microeconomics in a spatial interaction framework. **The major aim of this book is to make a more comprehensive effort to enhance spatial interaction models to share some of the essential properties of microeconomic models, whilst retaining their essential probabilistic character.** Whereas the derivation of spatial interaction models arises from a distinct framework projecting the behaviour of sets of actors at the macro level (eg. all firms of a certain type in a certain region) in terms of their perceived heterogeneity at the micro level (eg. each individual firm in a particular location), the *behavioural* constraints applied to the spatial interaction objective may arise from the fundamental relations of any theory appropriate to the analysis, such as microeconomic theory.

As an example, in the derivation of the regional supply functions for sets of competitive firms, the probabilistic model would evaluate these functions by maximizing the 'interaction' objective, whilst constraining the model to replicate the total value of the profits existing at the *observed* base period prices and quantities. The revenue would be expressed via a calibrated production function and the costs as the value of inputs purchased plus their transport costs. On the other hand, the classical deterministic model of the firm, as developed in microeconomics, would obtain the supply functions based on 'perfect' profit-maximizing behaviour. We also demonstrate that if the transformed objective functions of the probabilistic models are interpreted above as 'expected profits', they satisfy key economic criteria such as Hotelling's Lemma.

Whereas most conventional spatial interaction models have either total outflows constrained at the origins or total inflows constrained at the destinations or both, thus making the total supply or demand *exogenous*, microeconomic models regard both supply and demand as *endogenous* and price-responsive. In order to include such flexibility in spatial interaction modelling, it is necessary to distinguish between the constraints applied at the estimation phase of the model and those applied in the projection period. In the estimation phase, the constraints mainly add 'information' to the model by inducing its flows to conform with a wide range of observations, thus giving the modelled flows more chance of being close to the observed flows in the base period. However, except for say expenditure constraints on demand models, almost all of the estimation phase constraints are 'released' in the projection period, with their Lagrange multipliers being transformed into parameters. Thus, such projection models, enriched by the parameters associated with the constraints applied at the estimation phase, yield endogenous price-responsive values of all the unknown quantities.

When examining *regional* production capacities as distinct from the capacities of individual firms, it is clear that considerable heterogeneity usually exists within such aggregate capacities. This means that some sort of hierarchy of capacity efficiency will usually exist across a region. In the rare case where most of the capacity in a sector within a region is fairly homogeneous, the short run models merely introduce 'less than or equal to' constraints on capacity. However, when capacity is more heterogeneous across a region, we develop model structures to allow spill-over effects to adjacent regions as the preferred options in a more attractive region start to be exhausted. The resulting regional functions are of 'logistic' form, as argued intuitively by Hotelling over sixty years ago.

Spatial interaction models are generally specified as *discrete choice* models, where a housing region, an employment region, an intermediate input commodity from a given region, a travel mode or a transport link are regarded as the choice units. On the other hand, in microeconomics, the quantities demanded or supplied are treated as *continuous* variables. This distinction is especially important for models of retail and service consumption, where we need to distinguish the trips, which are discrete entities, from the quantities of different goods and services consumed per trip, which are continuous variables. Such a distinction allows both the trip frequency and the amounts of goods/services consumed per trip by each individual to be endogenous, which is vital for evaluating the travel consequences of both pricing policies and proposed facility and transport infrastructure investments. This book introduces some integrated discrete/continuous model formulations to handle these challenges.

Whereas the above discussion may imply that spatial interaction modelling is here being made 'subservient' to microeconomics, the book will also introduce some further concepts from geography, which are only recently starting to be applied in practical modelling. The concept of a *space-time profile* of the daily activity of members of households was introduced over twenty five years ago. This was at a time when the activity patterns of different members of a household were more rigidly defined than today, with, for instance, the mainly non-working spouses doing most of the shopping and service duties from home. More recently, the daily activity and associated travel patterns within households are becoming increasingly complex and interdependent, with both partners being employed and sharing tasks in different ways. Together with the recent 'self-imposed' increases in working hours in many professional disciplines, leisure time spent within the household is being increasingly prized, generating an increase in tele-commuting, as well as in tele-shopping and Internet shopping. In this book, an attempt is made to come to terms with these new trends by segmenting the households into income/life cycle groups and applying not only expenditure budget constraints, but also time budget constraints to households in each group. In addition, single-stop, single-purpose trips are generalized to multi-stop, multi-purpose tours on round-trip circuits.

Chapter Summaries In preparing a book to cover the above topics in an orderly way, three main sections have been formed. The first contains the 'Context and Methodological Framework'. Within this, we initially examine the history of spatial interaction modelling, mainly concentrating on the post-war period. As the book primarily tries to introduce microeconomics into spatial interaction modelling, the converse approach of the addition of 'space' into microeconomics forms the subject of the second chapter. Then using insights from both chapters, the *key third chapter* sets the principles and structure of a more unified spatial interaction modelling framework, in which the resulting models have properties generally consistent with the principles of microeconomics.

Becoming more specific, the second main section is entitled 'Some Relevant Urban Modelling'. With the fundamental recognition that travel is a demand 'derived' from the need to carry out activities at various times at other points in space, Chap. 4 looks at these activity decisions as the primary generators of both physical trips (tours) and tele-orders, mapping out the space-time profiles and the associated amounts of goods/services and travel consumed during the various daily tours. A companion supply model is formulated, with equilibrium and disequilibrium adjustments being examined under both perfect and imperfect competition. The open and efficient exchange of knowledge in a flexible multi-disciplinary environment is becoming a major identifier of successful cities. Chap. 5 looks at the importance of face-to-face contacts in this exchange process and makes a rather elementary effort to develop models to evaluate the resulting agglomeration economies. The supply of land for housing is a hierarchical but interdependent process, with broadacres purchased from farmers or government, lots and services developed on this released land and housing finally built on the lots. Chap. 6 formulates a three-level modelling system for these actors, looking at perfect and imperfect competition, questions of unsold stock and the handling of capacity constraints in the short run models.

In an attempt to be comprehensive, the last section covers 'Some Interregional Contributions'. Since its inception, an important application of spatial interaction models has been determining the flow patterns of commodities on transport networks. In Chap. 7, this topic is considered both in terms of changes in the transport networks themselves and changes in production technology, with different parts of the model structure being either deterministic or probabilistic. Chap. 8 presents an approach for evaluating interregional (or international) trade under imperfect competition, which is of increasing relevance for most sectors of the economy. In looking at interregional migration, the re-structuring of a region can best be understood by modelling 'gross' rather than 'net' migration. Thus, in Chap. 9, a probabilistic model of gross migration is developed, followed by a sequential dynamic procedure for coping with the inevitable leads and lags in the regional equilibration process.

A few numerical exercises are sprinkled throughout the two introductory chapters, all being capable of solution using a hand calculator. The solutions are grouped together at the end of each chapter.

The 'Future Challenges' section relates partly to an improved integration of the typically short run spatial interaction models with long run locational models, inevitably introducing the consideration of dynamics, as well as accounting for economies of agglomeration. In addition, the issue is addressed of developing better diagnostic statistics for these models, building on the well-known correspondence with maximum likelihood models, as well as recent work linking spatial autocorrelation and neural networks with spatial interaction.

Part I
Context and Methodology

1 Geographical Roots of Spatial Interaction

1.1 Analogies with Newtonian Gravitation

1.1.1 Gravity Effect Identification

Strong anecdotal evidence exists throughout recorded history for the operation of a 'gravity' effect in human activities. This is illustrated by the paramount role of accessibility in location decisions. In ancient times, major cities concentrated in seaport locations or along large rivers to be more accessible to trade opportunities with other cities. Also, export and the associated transport of locally produced commodities usually only took place after local demand was satisfied (except when there was an intrinsic price or quality advantage). Also, when faster and more reliable transport modes such as rail were introduced, major railheads increased in relative importance compared with river ports. In fact, most large river ports made sure that they also became important railheads, with some goods being handled by rail and bulk commodities by river barge.

At the urban level, the medieval cities were very compact, not only to reduce the perimeter for better defence against outside attack, but also to allow efficient pedestrian access to the available activities. For rural towns and large villages, where a high proportion of the population worked on encircling farms, a trade-off existed between a larger centre with better defence capabilities and smaller centres with less average distance to the fields and farming plots. Although little is known about the relative importance of these two countervailing influences, the newer cities started to spread out as civil peace became more widespread and as relatively cheap public transport became available from the 19th century onwards. Then, especially in the 'New World', further spreading at lower densities occurred in the 20th century, when public transport and car travel became increasingly affordable to members of the growing middle class.

For the medieval working class, who usually toiled from 'dawn to dusk', the time spent walking to their trades or farms was time subtracted from productive activity. By the middle of the 19th century, as fairly cheap public transport became available in the cities, travel *costs* (fares) were traded off against the *time* saved compared to walking. This illustrates the dual influence of cost and time of travel, which lies at the heart of the process of spatial interaction.

Newton's Theory of Gravitation reigned unchallenged in the latter part of the 19th century, before the appearance of Einstein's radical theories. In fact, the late

19th century has often been characterised as the period of highest optimism about the potential of science to explain everything observable in the universe. Thus, it is hardly surprising that suggestions started to arise for using Newton's theory to explain certain types of human activity between entities separated in space. If d_{ij} is defined as the distance between bodies i and j, m_i and m_j are the masses of bodies i and j respectively and k is a constant, then Newton's theory yields the gravitational force F_{ij} between bodies i and j as

$$F_{ij} = k\, m_i\, m_j\, (d_{ij})^{-2} \tag{1.1}$$

When applying Newton's theory to determine interaction between human activities at the urban or regional level, the gravitational force was replaced by the intensity of the interaction between the two areas, expressed as either the number of trips or moves between the areas. Also, the early researchers used directly the centroid-to-centroid distance between the interacting areas in place of 'd' above. However, the masses needed to be defined according to the type of activities being modelled. In one of the earliest works (Young 1924), which addressed migration between regions, the masses were merely defined as 'attraction' effects, without a clear physical interpretation. It was not until the work of Stewart (1941) that the masses were specified as the populations of the origin regions and destination regions of the movers. Although Stewart applied differential 'weights' to the populations in later work, he retained Newton's squared distance effect. A more empirical determination of a 'gravity parameter' awaited the efforts of retail modellers in the 60's.

1.1.2 Retail Gravity Models

The first serious effort to use gravity concepts to identify market areas for retailing was by Reilly (1929). He examined a typical intermediate town located between a pair of retail centres 'a' and 'b' separated by a distance 'd', where the town is a distance d_a from centre a and d_b from centre b, where $d_a + d_b = d$. If the size (eg. quantity of floorspace) of centres a and b is W_a and W_b respectively, the ratio of the proportions T_a and T_b of the total business attracted to each centre is

$$(T_a / T_b) = (W_a / W_b)\ (d_b / d_a)^2 \tag{1.2}$$

This implies that the business attracted to a centre from an intermediate town is directly proportional to the floorspace and inversely proportional to the square of the distance travelled between the town and the centre. From (1.2) it is also possible to compute at what distance d_b^* from centre b an intermediate town should be located such that the trade generated at centres a and b are equal. This result can be obtained by the setting of $T_a = T_b$ in (1.2) and transposing, yielding

$$d_b^* = d / [1 + ((W_a/W_b)] \tag{1.3}$$

The distance d_b^* was called the 'break even point' between centres a and b.

The above result is restricted to examining 'linear' pair-wise competition between retail centres with respect to one intermediate town. Also, the trade generated was only available as a *ratio* to that in the competing centre, rather than being known absolutely. With the growth of large regional shopping centres in major cities in the 60's and beyond, competition between the multiple regional centres themselves, as well as between the regional centres and the Central Business District (CBD) needed to be accounted for. In 1962, Huff, working in the Real Estate Research Program at the University of California, Los Angeles, rose to the challenge by developing a probabilistic retail model, which was published externally in Huff (1964). His model shifted the focus from the competitive situation of the retailer to the choice options available to the shopper, which in the end determine the (overlapping) trading areas associated with each of the centres and their resulting viability. The probabilistic structure recognises that a typical shopper in a large urban area does not shop exclusively at one centre, but chooses among alternative centres. More importantly, Huff moved away from the rigidity of the Newtonian model by (i) treating travel time rather than distance as the separation component and (ii) allowing the gravity exponent to be calibrated from observations, rather than being set *à priori* as *(-2)*. Thus, letting p_{ij} be the probability of a shopper in origin zone i shopping at centre j, t_{ij} the travel time between i and j and $(-\lambda)$ the gravity exponent, Huff's model is expressed as

$$p_{ij} = W_j\, t_{ij}^{-\lambda} \,/\, [\Sigma_j\, W_j\, t_{ij}^{-\lambda}] \qquad (1.4)$$

where Σ_j denotes the sum over all the competing centres j. In general, W_j denotes a measure of the relative *attractiveness* of destination j, here taken as linearly proportional to the floorspace. Finally, if O_i denotes the number of customers (trip origins) in zone i, then the number of trips S_{ij} between i and j is simply given as

$$S_{ij} = O_i\, p_{ij} \qquad (1.5)$$

From this pattern of shopping trips and assumptions on expenditure per trip, it is possible to evaluate expected turnover levels, parking requirements and customer catchment contours for each centre. In practice, the households can be classified into different income groups and the retail goods into categories, with a separate calibration being performed for each income group/good combination. In many earlier studies, shopping goods were just put into two categories, convenience goods (e.g. food) and comparison goods (e.g. furniture), with higher gravity exponents being obtained for the less durable goods. Note that, by the recognition of the gravity exponent $(-\lambda)$ as a parameter to be calibrated rather than an assumed constant, the Huff model must supplement Census data by surveys on the choice of the different centres by customers from the different origin zones.

A further interesting paper to appear in the 60's was Lakshmanan and Hansen (1966). Whereas it returned to distance rather than travel time as the gravity impedance measure, it defined the flows in terms of dollars rather than in terms of shoppers. This latter definition started to be questioned as gravity models became more *behavioural* from the end of the 60's – shoppers make the decisions to shop, not anonymous dollars! However, the paper did make one major contribution. Whilst the model, as for that of Huff, concentrated on modelling customer trip and expenditure patterns, it also examined the long term stability of the location and facility size distribution. For each class of goods, it assessed minimum levels of turnover per square foot of floorspace as profitability thresholds, and plotted these levels for the major centres in the Baltimore region. Also, in projecting likely future patterns of facilities, it gave greatest credence to those having relatively small variations in the expected turnover per unit area. In fact, this work anticipated in a heuristic sense the more formal locational equilibrium framework introduced in the path-breaking paper of Harris and Wilson (1978). In order to aid the understanding of the retail models and concepts of locational equilibrium, a small numerical example is provided below (see end of chapter for solution).

Exercise 1.1 Consider two adjacent small towns, where the first (larger) town is taken to generate $O_1=6000$ retail trips per day and the second town $O_2=4000$ retail trips per day. There is no difference between average incomes in the two towns. The smaller town has a somewhat lower residential density than the larger town. A retail centre is located at the centroid of each town, distances are treated as a proxy for travel time and a gravity parameter of 2 is chosen. With S_{ij} defined as the number of trips between town i and centre j, the Huff model becomes

$$S_{ij} = O_i W_j (d_{ij})^{-2} / [\Sigma_j W_j (d_{ij})^{-2}] \tag{1.6}$$

Assuming that $d_{11}=2$, $d_{22}=1.5$ and $d_{12}=d_{21}=5$, consider that 10000 floorspace units are required to service the entire region. Proceed as follows :

(i) Assuming initially that the current floorspace existing in each town is proportional to the trips generated from there, that is $W_1=6000$ and $W_2=4000$, compute the trip pattern S_{ij}^0 from the above model.

(ii) Compute the number of trips $S_j=\Sigma_j S_{ij}$ which are directed at each centre j. If unit costs and prices are taken as identical for each centre, the long run locational equilibrium will occur when the number of trips S_j is proportional to the floorspace W_j (or *equal to* W_j in our special case when $S_1+S_2=W_1+W_2$). This implies an equal profit per unit floorspace accruing to each centre, thus yielding no incentive for any of the small individual retailers within either of the centres to change location. Solve for this equilibrium pattern of floorspace and the associated flow pattern. This involves a simple iterative solution of a fixed point problem. Set $W_j^1=\Sigma_j S_{ij}^0$ where the S_{ij}^0 represent the flows evaluated in step (i). With these new values of floorspace, evaluate the new S_{ij}^1 values from the gravity model and re-evaluate the floorspace values $W_j^2=\Sigma_j S_{ij}^1$, and so on until convergence to the nearest unit of flow.

The above exercise, as well as introducing the concept of locational equilibrium as implied in Harris and Wilson (1978), also indicates the trade-off between size of centre (as a proxy for variety of goods available) and access distance. Thus, for the smaller town, the average (intrazonal) distance to its own shops is shorter than that for the larger town, yielding a higher *à priori* incentive to shop locally. However, this influence is in opposition to the potential benefits of shopping at the more attractive larger centre in the adjacent town. Characterisation of this trade-off is one of the most fundamental tasks of spatial interaction models. Note that, the above models all take attractiveness as directly proportional to floorspace. In later work (see Chap. 4), the attractiveness was taken as $(W_j)^\alpha$, where α can be regarded as an agglomeration index (from the standpoint of the customers) when greater than unity, and the reverse otherwise. In practice, α often turns out to be less than unity, indicating some level of redundancy in the largest centres. This parameter is typically evaluated using the method of maximum likelihood.

1.1.3 Gravity and Commodity Flows

Some of the earliest work on modelling trade flows was performed by Ohlin (1933). For any given commodity flowing between any two regions or countries, let X_r be the production of the commodity in origin region r, Z_s the consumption of the commodity in destination region s and $X\ (\equiv Z)$ the total production (or total consumption) of the commodity over all regions. Then, Ohlin's model gives the trade flows x_{rs} between any two regions r and s in the trading system simply via

$$X_{rs} = X_r Z_s / X \tag{1.7}$$

If in the basic Newtonian gravity equation (1.1) the masses are taken as X_r and Z_s and the constant k as $(1/X)$, then the Ohlin model emerges just for the special case when the gravity exponent is zero. In other words, the Ohlin model assumes that transport flows are independent of the distances through which the goods are shipped. They are merely jointly proportional to the production of the commodity at the origins and its consumption at the destinations. Although at first sight this model may appear to be suitable for goods where transport costs are quite a small proportion of their total value, such as high value-added goods, these goods are being transported increasingly by faster, more expensive modes of transport, including air. Thus, more advanced models must be sought where there is also an attempt to consider the influences of transport costs on flows.

A natural option would be to turn to a gravity model in terms of the distances d_{rs} between the regions. This would yield (1.7) in the form

$$X_{rs} = X\ \{X_r Z_s d_{rs}^{-n} / [\Sigma_{rs} X_r Z\ d_{rs}^{-n}]\} \tag{1.8}$$

which reduces to (1.7) when the gravity index n is zero. The denominator results from ensuring satisfaction of the total flow conservation condition $\Sigma_{rs}\ x_{rs} = X$.

Although this simple gravity model was used for some early studies, it was soon realised that the product of production potential, consumption potential and gravity influences did not explain some trade patterns very well. For instance, many countries have historical patterns of trade based on cultural affinities and perceptions of reliability in the timing and quality of deliveries. This problem was addressed by Theil (1967), where x_{rs} in (1.7) was multiplied by a bias factor Q_{rs}

$$Q_{rs} = (\bar{x}_{rs}\ \bar{X})/(\bar{X}_r\ \bar{Z}_s) \tag{1.9}$$

where the bars refer to the corresponding quantities above as observed at the base period. It is clear that such a model will perform well when mainly quantity changes occur between the base period and the projection period. However, when one is trying to determine the influence of expected changes in transport costs or import tariffs on future trade patterns, the use of (1.9) would be inappropriate. In practice, the distance terms in (1.8) should be replaced by corresponding transport cost terms, which would include the import tariffs.

Most commodities are used as inputs for the production of other commodities or finished goods. Thus, flows of a single commodity cannot be properly considered in isolation from the conditions of supply, demand and levels of technology in competing regions for all goods in the economy. Such issues will be discussed when spatial economics methods are introduced in Chap. 2.

1.2 Bases in Entropy, Constrained Optimization and Time Geography

1.2.1 Entropy Basis for Gravity Models

Philosophically, there are two distinct ways of expressing 'entropy'. Although the potential 'equivalence' of these two approaches has been well known in physics for over fifty years, the author and his colleague in Roy and Batten (1985) were among the first to make this connection in a very general way in regional science.

The first procedure, introduced to spatial modelling by Wilson (1967, 1970), arises from the field of statistical mechanics, which reflects an *enumeration* approach to statistics. The theory can be regarded as a macro theory with micro foundations. The most probable macro distribution is that which can be replicated by the maximum number of micro level events. In other words, the most probable macro allocation of a given activity, such as work trips, between a set of zones is *à priori* that which can be matched by the maximum number of potential trips between each individual worker in each zone of residence and each filled job in each zone of employment. The key assumption is that **each micro level event associated with a given macro distribution is equi-probable**. We note that this theory is applicable to any field where data availability restricts analysis to a macro level, but where identifiable discrete objects exist at the micro level, and is certainly not intrinsically a part of physics. *In fact, the approach is used in many contexts where (i) the number of individual objects is extremely large, (ii) data is* *only available at an aggregate level, (iii) one is only interested in having model results at this same aggregate level and (iv) the within-aggregate variance of the individual behaviour being modelled is much less than the between-aggregate variance. In cases where condition (iv) is unlikely to hold, one should perform market segmentation and carry out a separate analysis for each segment.*

A different concept of entropy was introduced by Shannon (1948) in terms of a representation of the uncertainty of a probability distribution, in which the most probable distribution is defined as that having the maximum entropy or uncertainty. Note that, this specification allows for distributions to be defined either with respect to macro behaviour by sets of individuals (eg. the probability of any randomly chosen worker commuting from a given residential zone and arriving at a given employment zone), or with respect to different discrete choices by specific individuals (eg. his or her choice of travel mode for the work journey). For example, the entropy S associated with the probability p_{hm} that a sampled individual h will choose mode m for the work journey is

$$S = - \Sigma_{hm} \, p_{hm} \, \log p_{hm} \tag{1.10}$$

The possibility of using entropy for individual discrete choice was introduced into regional science by Anas (1983). In most of the following, entropy based on macro/micro events via statistical mechanics will be applied. However, at several points the 'equivalent' probabilistic interpretation will be given to provide additional understanding at both the macro and individual levels.

Following Wilson (1967) and his successors, a *statistical mechanics* entropy formulation will now be presented for journey-to-work travel. At a certain base period let there be O_i workers observed in origin zone i and D_j filled jobs observed in destination zone j, yielding $T = \Sigma_i O_i = \Sigma_j D_j$ round trips in any one day. Let T_{ij} be the unknown number of trips between zones i and j. Also, let c_{ij} be the average generalised cost of travel (i.e. the money cost plus the time cost obtained with an assumed value of travel time) between zones i and j. From a base period survey, let the average generalised cost of travel over the entire urban area be c^0. Now, the objective is to find the 'macrostate' trip distribution T_{ij} at the base period which maximizes the number of associated micro level events (called *microstates*), consistent with satisfying all of the above base period observations, which are introduced via constraints which the model must replicate. These constraints are taken to add *'information'* to the model.

Now, the main principle for proceeding from macrostates to sets of associated microstates, as given by Roy and Lesse (1981) for regional science, is :

Individual objects only define microstates upon descending from any macro (aggregate) level to the individual object level when they are non-homogeneous (that is, distinguishable) with respect to the macro level behaviour being modelled (e.g. when modelling interzonal journey-to-work travel, commuters travelling between zonal pairs have different commuting time vs. job benefit trade-offs).

In addition, the individual jobs themselves within each zone may usually be of variable intrinsic attractiveness. Thus, the microstates are defined as each *distinguishable* worker within a set in a given origin zone which travel to a set of jobs in a given destination zone and occupy a *distinguishable* job there, taken over all origin-destination pairs. Using an improvement introduced by Cesario (1973), the entropy Z associated with a certain macrostate distribution T_{ij} is the number of ways that O_i distinguishable workers may be allocated from residential zones i to job zones j in groups T_{ij} times the number of ways that the $D_j = \Sigma_i T_{ij}$ distinguishable workers arriving at j may be allocated among the D_j distinguishable jobs there. Note that, when unemployment or unfilled vacancies exist, some prior steps need to be added to allocate the total workforce to those who are employed and total jobs to those which are filled (Roy, 1990). From permutation and combination theory, the number of microstates is given as

$$Z = \{ [\pi_i O_i! / \{\pi_j T_{ij}! \}] [\pi_j D_j!] \} \tag{1.11}$$

where π represents the product sign and $!$ the factorial. The first term represents the number of ways (combinations) that O_i distinguishable workers in each origin zone i may be allocated into destination sets $\{T_{ij}\}$. With the second term, we have the number of ways (permutations) that the D_j distinguishable workers arriving in zone j may be allocated into the D_j distinguishable jobs there, with only one worker allocated to each job. At this point, we introduce a small numerical example to illustrate the microstates.

Example *Consider a 2-zone system with two workers resident in each of the 2 zones and 1 job in zone 1 and 3 jobs in zone 2. That is,* $O_1 = O_2 = 2$, $D_1 = 1$ *and* $D_2 = 3$. *Let the observed macrostate be defined by the set of flows* $T_{11} = 1$, $T_{12} = 1$, $T_{21} = 0$ *and* $T_{22} = 2$. *Determine and illustrate the number of associated microstates, where each (m,n) pair denotes an individual worker 'm' from origin zone 'n'.*

Table 1.1 Allocation of (worker, origin zone) pairs to individual jobs

Job *1*, Zone *1*	Job *1*, Zone 2	Job 2, Zone 2	Job *3*, Zone 2
(1,1)	(2,1)	(1,2)	(2,2)
(1,1)	(2,1)	(2,2)	(1,2)
(1,1)	(1,2)	(2,1)	(2,2)
(1,1)	(1,2)	(2,2)	(2,1)
(1,1)	(2,2)	(2,1)	(1,2)
(1,1)	(2,2)	(1,2)	(2,1)
(2,1)	(1,1)	(1,2)	(2,2)
(2,1)	(1,1)	(2,2)	(1,2)
(2,1)	(1,2)	(1,1)	(2,2)
(2,1)	(1,2)	(2,2)	(1,1)
(2,1)	(2,2)	(1,1)	(1,2)
(2,1)	(2,2)	(1,2)	(1,1)

The microstates illustrate the allocation of *individual* workers from each origin zone to *individual* jobs in each destination zone, consistent with the T_{ij} distribution. For instance, with T_{21} being zero, only one of the workers from origin zone *1* can work in destination zone *1*. From (1.11) the total number of microstates is confirmed as *{ [2/{1*1}*2/{1*2}]* [1*6] } =12*, noting that *0!=1*.

Note that, the number of 'objects' in the above problem is only four. However, Wilson has demonstrated that the entropy microstate approach is only reliable when the number of objects per zone is very large. In this case, the distribution of Z (for the unconstrained problem) is very sharply peaked (Wilson, 1967), and the most probable result approaches the 'expected' solution. Wilson did not provide a companion proof for the more usual constrained problem [see Smith (1990) for this]. For practical purposes, the objective Z to be maximized can be replaced by the monotonic transformation $S=\log Z$, where the log is the *'natural'* log. Also, for cases with a large number of objects, the popular Stirling Approximation $x! = x(\log x - 1)$ can be applied. This enables (1.11) to be simplified to

$$S = \log Z = \Sigma_i O_i [\log O_i - 1] - \Sigma_{ij} T_{ij} [\log T_{ij} - 1] + \Sigma_j D_j [\log D_j - 1] \qquad (1.12)$$

In the solution of the entropy maximization problem at the base period, S is to be maximized in terms of the unknown (macrostate) set of trips $\{T_{ij}\}$. As the origin totals O_i and the destination totals D_j are known base period inputs to the problem, only the second term in (1.12) is involved in the differentiation.

Now the solution to the maximization of (1.12) is to be constrained by the model flows being induced to conform with certain aggregate base period quantities. Such constraints (except if they are strongly collinear) not only reduce the 'entropy' or 'uncertainty' of the final solution, but usually contribute to obtaining a $\{T_{ij}\}$ solution closer to that observed at the base period (if available). Such a model then becomes potentially viable for making projections for a future period. The constraints introduced by Wilson are first those inducing the row sums of the trips to all destinations to sum to the origin totals O_i in the form

$$\Sigma_j T_{ij} = O_i \qquad (1.13)$$

Secondly, we induce the column sums to equal the destination totals D_j in the form

$$\Sigma_i T_{ij} = D_j \qquad (1.14)$$

The above quantity sum constraints are generally known as the *accounting constraints*. Finally, as we are studying a particular class of behaviour, that is, of commuters, at least one *behavioural constraint* must be introduced. In Wilson's 1967 paper, this constraint reproduces the observed average generalised cost of travel c^0 in terms of the interzonal travel costs c_{ij}, as discussed earlier, yielding

$$\Sigma_{ij} T_{ij} c_{ij} = T c^0 \qquad (1.15)$$

Now if expression (1.12) is maximized in terms of T_{ij}, applying constraints (1.13) with Lagrange multipliers λ_i', constraints (1.14) with multipliers η_j' and constraint (1.15) with multiplier β, the solution emerges as

$$T_{ij} = exp - (\lambda_i' + \eta_j' + \beta c_{ij}) \tag{1.16}$$

As both $\Sigma_i O_i$ and $\Sigma_j D_j$ each sum to the total number of trips T, one of the equations in (1.13) or (1.14) is redundant, implying that the Lagrange multipliers λ_i' and η_j' are just known relatively, that is, they have one degree of indeterminacy. This means that (1.16) can be scaled by the product $(O_i D_j)$ of the respective row and column sums, scaling inversely to modify λ_i' to λ_i and η_j' to η_j. Now, denoting $exp - \lambda_i$ by A_i and $exp - \eta_j$ by B_j, (1.16) is finally given as

$$T_{ij} = A_i O_i B_j D_j \ exp - \beta c_{ij} \tag{1.17}$$

where the 'balancing factors' A_i and B_j are expressed recursively in the form

$$(A_i)^{-1} = \Sigma_j B_j D_j \, exp - \beta c_{ij} \quad ; \quad (B_j)^{-1} = \Sigma_i A_i O_i \, exp - \beta c_{ij} \tag{1.18}$$

This is the classical form of the *doubly-constrained* entropy model as introduced by Wilson and has also been come known as the trip distribution step in the 4-step transport planning model. The unknown balancing factors A_i and B_j are obtained by solving the recursive equations (1.18) iteratively.[1] Parameter β is usually evaluated by solving (1.15) via linear extrapolation. A good starting value β^0 usually turns out to lie between $(1.5/c^0)$ and $(2.0/c^0)$. In Wilson (1967), *origin-constrained* models without constraints (1.14) were introduced, as well as *destination-constrained* models without constraints (1.13).

Soon after Wilson's fundamental works, another important paper, (Karlqvist and Marksjö, 1971) appeared. The key achievement of this paper was to establish that if the solution of any entropy model, such as (1.17), has its unknown parameters (e.g. A_i, B_j, β) estimated by the method of maximum likelihood, then (i) the resulting likelihood equations correspond precisely with the constraints on the entropy model [e.g. (1.13) to (1.15)] and (ii) the estimated parameters are identical to the corresponding Lagrange multipliers of the entropy problem. This analogy has hardly been exploited in the literature, except for suggested use of maximum likelihood diagnostic statistics in spatial interaction models (Baxter, 1985).

As discussed in the prologue, there has been some ambiguity as to whether (1.17) should be taken as a model of work trips or as a model of job choice by workers in zones j who are resident in zones i. We maintain that when the trip costs c_{ij} correspond to those *revealed* in an actual trip pattern, then (1.17) just represents a model of work trips. However, if one would like to model the job choice process, the travel costs c_{ij} must be obtained from a *stated preference* survey, identifying the travel costs *expected* by all those deciding either to change job or enter the workforce during the study interval (Roy, 1989).

In making future projections where one has exogenous estimates of changes in the distribution of housing and jobs, as well as changed transport costs, the

[1] In the thorough treatment of gravity models in Sen and Smith (1995), a generic form resembling (1.17) is postulated, with chosen parameters estimated via maximum likelihood. Eq (1.17) does not embrace some alternative forms derived via microstates - see later.

'gravity impedance' multiplier β above is usually treated as a parameter, with the average trip cost emerging as an output. However, after each Census period, when new journey-to-work data becomes available, the model should be estimated anew. After several estimations over successive Census periods, one may clearly detect dynamic trends in the change of β, allowing good estimates to be made for β at a future time, rather than merely using the β from the most recent Census. In Chap. 3, the identification of further Lagrange multipliers of the estimated models as parameters for projection will be extended, building on Lesse (1982).

1.2.2 Intervening Opportunities Models

Before the above theoretical derivation of the gravity model by Wilson (1967), an alternative approach to modelling trip distribution was developed. This is the 'intervening opportunities' model, first developed by Stouffer (1940) and refined by Schneider for the Chicago Area Transportation Study (1959). This latter form, also described in Wilson's paper, is now presented for the journey-to-work.

For each origin zone i, let us rank possible destination zones j as j_{im} in order $\{m\}$ of their accessibility to zone i, where the 'closest' zone to i is denoted by $m=1$. Let P_{ijm} be the probability that one commuter will continue beyond zone j, the mth ordered destination out from zone i. Also, let there be a probability p that an available opportunity (job) will satisfy this commuter. If D_{ijm} is the number of jobs in the facility j ranked m^{th} out of zone i, then we can write

$$P_{ij1} = (1 - p\, D_{ij1}) \qquad (1.19)$$

Assuming independence, the probabilities are treated multiplicatively upon passing each successive destination, leading to the recurrence relations

$$P_{ijm} = P_{ij\,m-1}\,(1 - p\, D_{ijm}) \qquad (1.20)$$

If (1.20) is transposed in terms of D_{ijm}, and we define W_{ijm} as the total number of jobs 'passed up' to and including j_{im}, we can write

$$D_{ijm} = W_{ijm} - W_{ij\,m-1} \qquad (1.21)$$

enabling (1.20) to be expressed as

$$(P_{ijm} - P_{ij\,m-1}) / P_{ij\,m-1} = -p\,(W_{ijm} - W_{ij\,m-1}) \qquad (1.22)$$

Assuming a continuous transition, this yields the incremental relations

$$dP / P = -p\, dW \qquad (1.23)$$

This is an elementary differential equation, having the solution

$$P_{ijm} = k_i \ exp - (p \ W_{ijm}) \tag{1.24}$$

where k_i is a constant. With O_i commuters being available at zone i, the total number T_{ijm} of trips stopping at the mth destination j out from zone i is

$$T_{ijm} = O_i \, (P_{ij,\, m-1} - P_{ijm}) \tag{1.25}$$

Upon substitution from (1.24), this yields

$$T_{ijm} = k_i \, O_i \, [\ exp - (p \ W_{ij\, m-1}) - exp - (p \ W_{ijm}) \ ] \tag{1.26}$$

Now the flows should still satisfy the origin constraints (1.13), which upon substitution from (1.26), with M the total number of zones, gives the constant k_i as

$$k_i = 1 \, / \, [1 - exp - (p \ W_{ijM})] \tag{1.27}$$

The above model has two major deficiencies (i) its inability to satisfy the destination constraints (1.14) and (ii) the fact that destinations can be distributed over 360 degrees surrounding the origin zone i, implying that the opportunities in successive destinations at increasing distance out from i may not be truly 'intervening'. In other words, successive destinations could lie as much as 180 degrees around from each other along axes out from zone i. Problem (ii) can be overcome by defining separate sets of intervening opportunities for each pair of origin zones i and destination zones j, rather than just for each origin zone i. Intervening opportunities would then be identified at all destinations on plausible paths connecting each origin zone i and each destination zone j. Compliance with the destination constraints (1.14) can be obtained by carrying out a further origin and destination balancing process (Wilson ,1967).

Although Wilson (1967) derived the above intervening opportunities model using an entropy microstate approach, it still had problems (i) and (ii) as described above. In the following, a combined gravity/intervening opportunities model for the work journey will be derived, following Gonçalves and Ulysséa-Neto (1993) and Roy (1993). The derivation follows that of the classical doubly-constrained gravity model in (1.17), with the addition of a constraint on the observed average number of intervening opportunities w (jobs) 'passed' per trip in the base period on the way to the final destination. As the first step, the most plausible single path connecting each origin zone i and each destination zone j in the system is identified, and for each such path the number of jobs w_{ij} passed upon crossing the various job zones on the way to the final destination zone j is summed. Although the following analysis only considers a single path between each origin zone and each final destination zone, it is possible to extend the approach to include multiple paths. As for the travel cost constraint (1.15), an intervening opportunities constraint is added in the form

$$\Sigma_{ij} \, T_{ij} \, w_{ij} = T \, w \tag{1.28}$$

It is intuitively clear that the Lagrange multiplier α on (1.28) will have the same sign as the multiplier β on the travel cost constraint (1.15). This is because a large number of intervening opportunities passed on the way to a final destination should decrease the *à priori* probability of choosing that destination. Thus, the combined model yields the conventional model (1.17) in the revised form

$$T_{ij} = A_i O_i B_j D_j \, exp - (\alpha w_{ij} + \beta c_{ij}) \qquad (1.29)$$

As discussed in Roy (1993), when there is a reasonably constant density of the intervening opportunities per unit distance between i and the alternative final destinations j, the α and β constraints may become partly 'collinear', introducing convergence problems into the linear extrapolation estimation procedure for determining these parameters. In such cases, the intervening opportunities constraint (1.28) should be replaced by a constraint on the average number of intervening opportunities which are passed *per unit distance*.

1.2.3 Activities, Time Budgets and Travel

Especially for discretionary travel, including trips for shopping and personal services, trips may be both *multi-purpose and multi-stop*. Also, many work trips now include lunch-time travel or stops on the way to or from work. The time taken for such travel clearly competes with leisure time spent at home. The field of *'activity analysis'*, introduced by Hägerstrand (1970), explicitly recognises this trade-off. He postulated that the behavioural assumptions underlying most of the urban and regional models of the time were still based on a crude representation of real spatial behaviour. In particular, spatial interdependencies were usually ignored, as well as interdependencies between households. These concepts started to influence model structure in important work at Oxford, such as in Jones (1976), where multi-stop and multi-purpose journeys were considered. Discrete choice theory contributions were introduced by Lerman (1979) and Kitamura (1984). O'Kelly (1981) adopted a Markov chain approach and Southworth (1985) used simulation combined with chain accessibility measures obtained by summing incremental accessibilities between successive stops. As Mazurkiewicz (1985) developed a *spatial interaction* framework based on *explicit* trip chains, his method is described below, concentrating on his 'alternative' approach[2]. Some of the other papers are described in Sect. 1.3.5, where the important concept of *'competing destinations'* is also described, as introduced by Fotheringham (1983a) to handle destination *'adjacency'* effects.

Mazurkiewicz defines a point where residences are concentrated, surrounded by concentric rings $k = 1,2, ..K$ of equal width, over which options are distributed at uniform density. A typical 'movement' M_k^s is described as a trip with s stops out from the origin, terminating in ring k and returning non-stop to the origin. This

[2] Mazurkiewicz thanks an anonymous referee for suggesting this alternative but closely 'equivalent' formulation. I was that referee.

means that between the origin and ring $(k-1)$, there are $(s-1)$ stops at intermediate rings. Assuming that travel out from the origin is radial, let 'x' be the travel distance between concentric rings. In evaluation of the total number Z of microstates in the system, consider first the number of ways that a total of M distinct movements may be distributed into sets of movements M_k^s. Secondly, each movement within the $(k-1)$ rings out from the origin is allocated arbitrarily to single stops at $(s-1)$ intermediate rings, with arbitrary repetition of identical stop patterns being admissible between any of the M_k^s journeys. Then, using permutation and combination theory, the number of microstates is given as

$$Z = \{ M! / [\pi_{ks} M_k^s !] \} \{ \pi_{ks} [(k-1)! / ((s-1)! (k-s)!)]^{M_k^s} \} \tag{1.30}$$

As constraints, there are the origin constraints

$$\Sigma_{ks} M_k^s = M \tag{1.31}$$

and the travel distance constraint

$$\Sigma_{ks} M_k^s k x = M \bar{x} \tag{1.32}$$

where $\bar{x}$ is the 'sampled' average distance travelled per complete movement.

Defining the entropy S as $\log Z$, applying the Stirling approximation (as before) to the first [...] term in (1.30), S can be expressed as

$$S = - \Sigma_{ks} M_k^s [\log M_k^s - 1] + \Sigma_{ks} M_k^s \log n_{ks} \tag{1.33}$$

where $n_{ks} = [(k-1)! / \{(s-1)! (k-s)!\}]$. Upon differentiation of (1.33) and attached constraints (1.31) and (1.32) with respect to M_k^s, as well as removal of the Lagrange multiplier associated with constraint (1.31), we obtain the solution

$$M_k^s = M\ n_{ks} \exp - \beta k x / [\Sigma_{ks} n_{ks} \exp - \beta k x] \tag{1.34}$$

where β is the gravity parameter, that is, the Lagrange multiplier associated with the distance constraint (1.32). From (1.34), it is seen that, as the index k of the final destination ring increases, n_{ks} increases, but $\exp - \beta k x$ decreases, illustrating a trade-off in trip frequency between the availability of intermediate rings for stops and the distance to the final ring k. This represents the most fundamental trade-off in activity analysis.

Exercise 1.2 As an illustration of the above model, consider one particular type of trip $M_4^3 = 2$ out from the origin, where with both travellers making their 3rd stop at the 4th ring, there are two more stops made at arbitrary rings of the three preceding rings. Let the two travellers be identified by 'a' and 'b' and let a typical microstate be written as : $a \to (1,2)$ $b \to (1,3)$, denoting 'a' making stops at rings 1 and 2 and 'b' making stops at rings 1 and 3.

(i) Assuming that the microstates are distinguished by different stops of both '*a*' and '*b*', enumerate the number of microstates associated with M_4^3 and confirm that they coincide with the second term in (1.29) as : $[\,(k-1)!\,/\,((s-1)!\,(k-s)!)\,]^{M_k^s}$

(ii) If the $(s-1)$ intermediate stops are now allowed to occur arbitrarily across the outward and inward journeys (for instance, 2 stops on the way out and 1 stop on the way home), distinguishing between stops at the same rings on the outward vs. the homeward trips, revise the expression for the microstates in (i) above to reflect this greater freedom of choice. Also, re-evaluate the new number of microstates.

Of course, the above model has made a strong simplification of zonal geometry and travel routes to arrive at a rational formulation. For each trip, the destinations are equally spaced. Also, the travel routes are purely radial from a single origin. In addition, the distribution of options is uniform across the circle and stops are assumed to occur at *all* intermediate rings. Furthermore, the M_k^s multi-stop movements are not of themselves important for planning purposes. In order to check the viability of the different service destinations, it is necessary to know the total number of stops S_k at each destination, as a measure of turnover. Although this was not mentioned in the original paper, S_k can be evaluated as

$$S_k \;=\; \Sigma_s M_k^s + \Sigma_{s,\, l=k+1,K}\; M_l^s\,(s-1)/(k-1) \qquad (1.35)$$

where the first term represents the stops at k for all trips terminating at k and the second the sum of all intermediate stops at k for trips proceeding beyond k and terminating at any ring between $k+1$ and K.

As the first paper to come to grips with the combinatorics of multi-stop trips, Mazurkiewicz (1985) is an important contribution, despite the limitations noted above. However, one key omission is consideration of any *time budget* constraint on the multi-stop trip pattern. The inclusion of time budget constraints requires a major expansion of the conventional spatial interaction framework in three ways. First, with the time being spent on travel plus at all stops on a trip chain being clearly dependent on trip frequency, it will be necessary to *endogenise* household trip frequency. Secondly, if it is thought desirable to introduce the time spent at any facility on a trip circuit as a *continuous* random variable, our analysis must be extended from just the modelling of *discrete* trips. And thirdly, a 'numeraire' good, such as 'other leisure time' will need to be included so that the time budget constraint is satisfied as an equality with a non-zero Lagrange multiplier, rather than as an inequality with a zero Lagrange multiplier, which can have no influence whatsoever on the results. All these considerations are discussed thoroughly in Chap. 4, where both time and money budget constraints are included, unifying the time geography and microeconomics approaches.

1.2.4 Entropy in Continuous Space

In the previous formulations, all origin activities and all destination activities were taken as concentrated at a single point within the encompassing zones. Although this may be a reasonable approximation for destination facilities such as shopping centres and office blocks, it is a poor representation of the spatial distribution of households, which are more appropriately depicted as a continuous distribution over each zone. These issues were addressed by Angel and Hyman (1972) in developing a continuous space form of Wilson's doubly-constrained gravity model. Their formulation utilised some of their earlier ideas on urban velocity fields (Angel and Hyman, 1970), where they obtained *area* measures of congestion rather than the usual *network* measures. Travel in continuous space was also treated by Puu (1979). However, as his approach was not cast in the form of a maximization problem, it is less helpful for illustrating the analogies with the conventional discrete gravity model than the Angel and Hyman formulation. Thus, only the latter is presented in the following discussion, drawing on concepts in both their 1970 and 1972 papers.

In this analysis, neither a transport network nor a zoning system is considered. In place of the discrete quantities of the conventional model, we have continuous distributions of residences, workplaces and travel time (or cost) between any two locations. The geometry of the city is represented by a polar coordinate system, with coordinates (r_1, θ_1) representing a point of trip origins and (r_2, θ_2) denoting a point of trip destinations. An origin density function $O(r_1, \theta_1)$ is defined as the trips originating per unit area about point (r_1, θ_1) and a destination density function $D(r_2, \theta_2)$ as trips per unit area coming into point (r_2, θ_2). Then, the trip density function $T(r_1, \theta_1, r_2, \theta_2)$ represents the number of trips between a unit area surrounding point (r_1, θ_1) and a unit area surrounding point (r_2, θ_2). Finally, from the 'area' approach to congestion described in Angel and Hyman (1972), a travel time function $t(r_1, \theta_1, r_2, \theta_2)$ is determined between any points (r_1, θ_1) and (r_2, θ_2). By analogy with the discrete case, the entropy S is expressed as

$$S = -\int_0^{\infty}\int_0^{\infty}\int_0^{2\pi}\int_0^{2\pi} T(r_1, \theta_1, r_2, \theta_2)\,[\,\log T(r_1, \theta_1, r_2, \theta_2) - 1]\, r_1\, r_2\, d\theta_1\, d\theta_2\, dr_1\, dr_2 \tag{1.36}$$

where integration is occurs outwards from the city centre to a radius → infinity[3].

Considering the constraints, at all points (r_1, θ_1) we must satisfy

$$\int_0^{\infty}\int_0^{2\pi} T(r_1, \theta_1, r_2, \theta_2)\, r_2\, d\theta_2\, dr_2 = O(r_1, \theta_1) \tag{1.37}$$

and at all points (r_2, θ_2)

$$\int_0^{\infty}\int_0^{2\pi} T(r_1, \theta_1, r_2, \theta_2)\, r_1\, d\theta_1\, dr_1 = D(r_2, \theta_2) \tag{1.38}$$

analogous to the origin and destination constraints (1.13) and (1.14) respectively

[3] In an actual city, we can use the true outermost radius as the upper integration limit.

in the 'corresponding' discrete model. As the counterpart to the travel cost (time) constraint (1.15) of the discrete model, we have here

$$\int_0^{\infty}\int_0^{\infty}\int_0^{2\pi}\int_0^{2\pi} T(r_1,\theta_1,r_2,\theta_2)\; t(r_1,\theta_1,r_2,\theta_2)\; r_1 r_2\, d\theta_1\, d\theta_2\, dr_1\, dr_2 = T\,\bar{t} \tag{1.39}$$

where $\bar{t}$ is the base period average travel time per commuting trip over the whole system. The origin and destination trip densities must each yield the total number of trips T when integrated over the complete circle. If the objective (1.36) is maximized with respect to the unknown trip densities under constraints (1.37) to (1.39), using differentiation under the integral sign, an elegant functional form is obtained for the trip densities as follows:

$$T(r_1,\theta_1,r_2,\theta_2) = A(r_1,\theta_1)\, O(r_1,\theta_1)\, B(r_2,\theta_2)\, D(r_2,\theta_2) \exp -\beta\, t(r_1,\theta_1,r_2,\theta_2) \tag{1.40}$$

in analogy to (1.17) where the 'balancing factor' functions come out as

$$[A(r_1,\theta_1)]^{-1} = \int_0^{\infty}\int_0^{2\pi} B(r_2,\theta_2)\, D(r_2,\theta_2) \exp -\beta\, t(r_1,\theta_1,r_2,\theta_2)\, r_1\, d\theta_1\, dr_1 \tag{1.41}$$

for the origins and for the destinations, in analogy to (1.18), as

$$[B(r_2,\theta_2)]^{-1} = \int_0^{\infty}\int_0^{2\pi} A(r_1,\theta_1)\, O(r_1,\theta_1) \exp -\beta\, t(r_1,\theta_1,r_2,\theta_2)\, r_2\, d\theta_2\, dr_2 \tag{1.42}$$

Unless the integrals in (1.41), (1.42) and (1.39) with substitution of (1.40) have analytical solutions, the balancing factor functions and the gravity parameter must be evaluated by numerical integration techniques. If radial symmetry is assumed, the equations simplify and analytical solutions may emerge. Also, the velocity field computations (Angel and Hyman, 1970) to obtain the travel time distribution $t(r_1, \theta_1, r_2, \theta_2)$ are made before the above trip distribution step is carried out. This is highly unusual, as the trip distribution step is conventionally followed by trip assignment to load the trips onto the transport network. Thus, whilst the approach is conceptually interesting, it may be difficult to apply in practical situations.

One of the most sensitive parts of the gravity model is the treatment of intrazonal travel, where both the worker and his or her job lie in the same zone. With travel time represented inversely in the gravity model, the low intrazonal travel times have a strong influence on the overall solution. This situation has been identified in the literature as the *self potential* problem. The main difficulty is to determine the average trip time for intrazonal travel. Some key references in this area include Bonsall (1975), Bröcker (1989) and Frost and Spence (1995). It is also clear that interzonal travel times may also be in error, being based on centroid-to-centroid distances between the zones. This error can be especially significant when the job zone of a worker is contiguous to his or her zone of residence. Thus, in Roy (1999), some of the ideas from the above Angel and Hyman (1972) analysis were used to simulate an urban system comprised of circular abutting zones in which radial/circumferential travel was considered on shortest path routes. This analysis, recognising continuous geometry within each zone, allowed the relative magnitude of both intrazonal errors and interzonal errors to be assessed. In practice, urban areas contain zones of quite irregular shape, making the use of continuous geometry very cumbersome. Instead, the work in Roy (1999) was directed at developing 'rules of thumb' to correct a conventional discrete zone analysis for the effects of continuous geometry. The paper revealed errors in flows of 15% and more in some parts of the discrete model, especially for high values of β. The ideas of Angel and Hyman were also used in Roy (1997a), where integration was over continuous quantities of goods and services consumed per trip rather than continuous space, yielding consistency with microeconomic demand theory (see Chap. 4 for more detail).

1.2.5 Further Entropies and Their Physical Significance

In physics, different types of entropy are used for different situations, depending on whether the objects being allocated or their destinations are *distinguishable* or not. For instance, the houses available for sale in a certain region are certainly distinguishable. On the other hand, beds in a public hospital, not being able to be chosen specifically by the patients, are not distinguishable from the viewpoint of patient demand. A further variation occurs with respect to the number of objects allocated into a particular discrete category. Whereas only one detached house may be allocated to a given allotment or one worker to a given job, an arbitrary number of shoppers may be allocated to say 100 m^2 of floorspace in a retail centre during a certain period. Fisk and Brown (1975), Lesse *et al.* (1978) and Roy and Lesse (1981) adapted entropies from physics to particular contexts in spatial planning. This work is briefly summarised below.

a. Maxwell-Boltzmann Entropy In this case, we consider the number of ways N that x distinguishable objects can be allocated into groups x_j, where $j = 1, J$. This form, used earlier for the doubly-constrained gravity model, is expressed as

$$N = x! / \pi_j x_j !$$

b. Fermi-Dirac Entropy When a particular slot or category is either *occupied* or *unoccupied*, that is, it has a $\{0, 1\}$ status, Fermi-Dirac entropy is appropriate. Consider n distinguishable locations with $\{0, 1\}$ occupancy. Then, N is the number of ways that x indistinguishable objects (where $x<n$) may be allocated into the n distinguishable locations, given as

$$N = n! / [x! (n-x)!]$$

c. Bose-Einstein Entropy If each of the n distinguishable locations may be either unoccupied or occupied by an arbitrary number within the total of the x indistinguishable objects, then N changes from the above to the following

$$N = (n + x - 1)! / [(n-1)! x!]$$

The above three distributions from physics turn out to be insufficient to define all relevant situations for planning. It is necessary to define two more useful forms from Lesse *et al.* (1978), which were given specific planning interpretations in Roy and Lesse (1981).

d. *{0, 1}* Occupancy and Both Objects and Locations Distinguishable This is a useful variation of Fermi-Dirac entropy. Let N be the number of ways that x distinguishable objects may be placed into n distinguishable locations $(x<n)$ with *{0, 1}* occupancy, given as

$$N = n!/(n-x)!$$

e. Arbitrary Occupancy and Both Objects and Locations Distinguishable In this case, we make a variation on Bose-Einstein entropy. Let N be the number of ways that x distinguishable objects may be placed into n distinguishable locations with arbitrary occupancy, yielding

$$N = n^x$$

Roy and Lesse (1981) used analogies with logit models derived from random utility theory to identify distinguishability in the 'corresponding' entropy models with the quantities associated with error terms in the utility models. In practice, most individual objects within categories in geography and economics are distinguishable. However, as mentioned by Fisk and Brown (1975), individual shipments of bulk commodities from the one producer can usually be regarded as indistinguishable, allowing Bose-Einstein entropy to be used. On the other hand, the shipments from different producers within a region may vary in quality, implying that the producers within a region should usually be treated as distinguishable. When using Fermi-Dirac entropy for multiple regions j, it is noted that when the total capacity $\sum_j n_j$ in the system is greater than the total number of objects $\sum_j x_j$ to be allocated (often associated with the demand), the model will ensure that $x_j < n_j$, that is, *capacity can never be exceeded in any single zone*. Thus, Fermi-Dirac entropy can be regarded as applying a logistic capacity constraint to each region as an alternative to a sharp inequality constraint. This useful property will be discussed in detail in chapter 3. An exercise follows, expanding on the example presented in Roy and Lesse (1981).

Exercise 1.3 Consider the problem of locating four households into two zones, where zone 1 has 2 vacant lots and zone 2 has 4 vacant lots. We have no further information. For each of the following assumptions on distinguishability and lot occupancy, determine the 'most probable' macrostate *(a,b)*, where '*a*' is the number of households allocated to zone *1* and '*b*' the number to zone *2*, as that yielding the maximum number of microstates Z, by checking all possibilities :

(i) Assuming that just the households are distinguishable and that multiple occupancy of lots is possible, use Maxwell-Boltzmann entropy a. above to find the most probable *(a,b)* pair, that is, yielding the most microstates [*Hint* We try all pairs *(0,4)*, *(1,3)*, *(2,2)*, *(3,1)*, and *(4,0)*].

(ii) Assuming that just the lots are distinguishable and that all the lots are single occupancy, use Fermi-Dirac entropy b. above to determine the most probable *(a,b)* pair [*Hint* We have the pairs *(0,4)*, *(1,3)* and *(2,2)*].

(iii) Assuming that just the lots are distinguishable and that multiple occupancy is allowable, use the Bose-Einstein entropy c. above to determine the most probable *(a,b)* pair.

(iv) Assuming that both households and lots are distinguishable and that all lots are single occupancy, combine the two entropies a. and d. above to find the most probable *(a,b)* pair.

(v) Assuming that both households and lots are distinguishable and that multiple occupancy is allowed, combine the two entropies a. and e. above to find the most probable *(a,b)* pair.

Whereas all the above entropies are expressed according to the statistical mechanics approach, some can also be interpreted in terms of the Shannon entropy of alternative probability distributions (Roy and Batten, 1985), as defined in (1.10). For instance, if p_j is defined as the probability that a household will be allocated to a lot in zone j, the Maxwell-Boltzmann entropy a. above is given in probabilistic form as

$$S = - \Sigma_j p_j \log p_j$$

On the other hand, Fermi-Dirac entropy b. is equivalent to a Shannon entropy using the probability p_{jk} that any lot chosen at random is both in zone j and in state k, where $k=0$ denotes it is vacant and $k=1$ indicates that it is occupied by one household, yielding

$$S = - \Sigma_j p_{jk} \log p_{jk}$$

In general terms, it is seen that the probabilistic entropy equivalent to Maxwell-Boltzmann is demand-oriented. On the other hand, the Fermi-Dirac equivalent above relates demand to supply capacity. These are the two most obvious examples. For the entropies c. to f., we can often relate the statistical mechanics entropy to limiting cases of probability distributions (Roy and Lesse, 1985).

In Evans (1973), a fundamental result was demonstrated which is valid for all forms of entropy. Using the particular case of the doubly-constrained entropy model of (1.12) to (1.16), it is shown that, as the gravity parameter β approaches infinity, the result approaches asymptotically that of minimizing the total travel cost on the right-hand side of (1.15) under the origin and destination constraints (1.13) and (1.14) respectively. The latter is known as the transportation problem of linear programming (LP). Thus, as illustrated in Fig. 1.1, there is a seamless transition, as β starts increasing towards infinity, between the solution of our probabilistic entropy model and the solution of the deterministic LP, the *objective* of which is the cost (behavioural) *constraint* of the entropy model. This result implies that the deterministic problem can always be obtained as a 'special case' of our entropy problem by making β sufficiently large. More importantly, we are free to construct probabilistic versions of classical deterministic models arising from different theories, such as the profit maximization model of identical perfectly competitive firms aggregated to the regional level arising from microeconomic theory. In this case, the cost constraint (1.15) is merely replaced by a profit constraint, with the revenue term expressed via a production function (which gives output as a function of the quantities of all the material and human inputs) and the cost term given as the sum of the costs of all these inputs and their transport costs to the places of production. This enhancement lies at the heart of

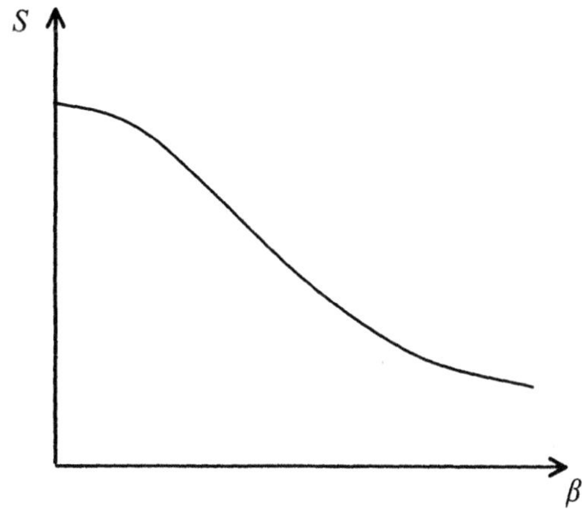

Fig. 1.1 Entropy *S* versus Gravity Parameter *β*

efforts to link probabilistic spatial interaction modelling with some of the deterministic models of microeconomics, one of the book's key aims.

1.3 Further Enhancements

1.3.1 The More General Information Theory

Information theory (Kullback, 1959) is a *statistical inference* technique which evaluates the change in a probability distribution resulting from the supply of certain new *information*. The *information content h(x)* of a message about an event that is expected to occur with probability x satisfies the following axioms:

- The information content $h(x)$ is taken as a decreasing function of the probability x, and is usually taken as $h(x) = log_e (1/x) = - log_e x$.
- $h(x)$ is a continuous function of x $(0<x<1)$
- $h(o) = \infty$ and $h(1) = 0$
- $h(x)$ is a monotonically decreasing function of x and
- Additivity of information exists for independent events, that is $h(x_1, x_2) = h(x_1, x_2)$.

From the above, the *information gain* of a message predicting that an event will occur according to an *à posteriori* probability x_1 with respect to an *à priori* probability x_0 is given as

$$I(x_1, x_0) = log_e (x_1/x_0) \tag{1.43}$$

It follows that if we have n events indexed by *i*, then given prior probabilities q_l and posterior probabilities p_i, then the information gain $I(p, q)$ is given as

$$I(p, q) = \Sigma_i p_i log (p_i / q_i) \tag{1.44}$$

where, for compactness, the subscript '*e*' has been omitted from the natural log. Finally, if a prior distribution q is available, as well as some further posterior information in the form of constraints, then the *most probable* posteriori distribution p is that which *minimizes* the information gain $I(p, q)$ in (1.44) above. This conservative hypothesis implies that one does not assume any more gain of information than one is entitled to expect from the extra constraint data. The reader may compare the Shannon entropy expression (1.10) with the information gain relation in (1.44). If the prior distribution q is the uniform distribution, (1.44) can be seen to reduce to an expression analogous to (1.10). In other words, Shannon entropy becomes 'equivalent' to information gain when one has no *à priori* information whatsoever, or when one has no trust in what one has. For instance, if the prior distribution reflects the events observed at a previous point in time, and if there has been a fundamental change in the system since that time, it may be inappropriate to use q as an *à priori* bias in characterising the behaviour of the current system.

Two key methods which have been used extensively in regional science were related to information theory in the key paper of Snickars and Weibull (1977). These are the RAS method (Cambridge University, 1963) for input-output analysis and the Cross-Fratar approach (Fratar, 1954). The latter approach is presented here, illustrated for the doubly-constrained trip distribution model.

From the entropy microstate derivation of this model in (1.12) to (1.18), we can define $p_{ij} = (T_{ij} / T)$ as the probability that any trip chosen at random occurs between zones i and j. If q_{ij} is defined as the observed probability of such an event at the (usually) most recent time period at which data is available (for example, at the latest Census), then the information gain objective (1.44) is generalised to this bivariate distribution as follows

$$I(p, q) = \Sigma_{ij} p_{ij} log (p_{ij} / q_{ij}) \tag{1.45}$$

If both sides of the origin and destination constraint relations (1.13) and (1.14) are converted into probabilistic form by dividing by the total number of current trips T, then minimization of (1.45) under these constraints yields [see (1.16)]

$$p_{ij} = q_{ij} exp - (\lambda_i^* + \eta_j^*) \tag{1.46}$$

where λ_i^* and η_j^* are the new Lagrange multipliers on the normalised (1.13) and (1.14) respectively. In comparison with (1.16), we note that q_{ij} replaces the gravity impedance effect $exp - \beta c_{ij}$. In a similar way as reasoned in proceeding to (1.17) and (1.18), the result (1.46) can be expressed in the form

$$p_{ij} = q_{ij}\, a_i\, o_i\, b_j\, d_j \tag{1.47}$$

with $o_i = O_i / T$, $d_j = D_j / T$ and a_i and b_j expressed recursively as

$$(a_i)^{-1} = \Sigma_j\, q_{ij}\, b_j\, d_j \quad ; \quad (b_j)^{-1} = \Sigma_i\, q_{ij}\, a_i\, o_i \tag{1.48}$$

As demonstrated by Snickars and Weibull (1977), the 'Fratar' method (1.47)-(1.48) usually performs better than the entropy gravity model (1.17)-(1.18). Of course, if one *happened* to know the average trip cost c for the projection time period and added this information via a normalised version of the constraint (1.15) to our information theory objective, they found that the result was still better than the straight Fratar method above. However, as the average trip cost c represents a key property of the *internal* state of the transport system being modelled, it will be an *output* from rather than an *input* to our projection model. Thus, *in practice*, we seem to be 'stuck with' either the Fratar model (1.47)-(1.48) or the doubly-constrained gravity model (1.17)-(1.18).

Although Fratar in (1.47)-(1.48) seemed to perform better than the gravity model, it has no explicit representation of transport costs. This means that when, for example, there is a major change in the transport system between the base period (corresponding to our observed 'prior' distribution) and the projection period, such as the opening of a large new section of freeway, the Fratar model above has no way of representing this change. On the other hand, the gravity model (1.17)-(1.18) merely enters the new generalised transport costs c_{ij}^* in place of the original costs c_{ij} and re-balances (1.18) to compute the projected flows. However, if, as seen above, it did not do a very good job in fitting to the observed base period flows, we can only expect reasonable accuracy in projecting the *relative* changes from the base period, rather than having a good estimate of the *actual future* flows. This conundrum was addressed in Roy (1987), where an approach was presented to combine the greater projection reliability of the Fratar procedure with the sensitivity to transport network and cost changes of the conventional gravity model. This approach is summarised below.

At the base period, let us solve the doubly-constrained gravity model (1.17)-(1.18) for both the modelled flows T_{ij} and the impedance parameter β. If an error measure such as RMS (Root Mean Square) evaluated between the modelled base period flows T_{ij} and the observed (eg. Census) base period flows T_{ij}^0 is found to be unacceptably high, then this gravity model cannot be considered as reliable for projection. This situation will quite often occur, as illustrated by Snickars and Weibull (1977) for Stockholm. In that case, the combined approach of Roy (1987) is recommended, as follows. Let the ratio (T_{ij}^0 / T_{ij}) of the observed base period flows and the base period flows modelled by the gravity model (1.17)-(1.18) be denoted by x_{ij}. Normalise this to represent *bias probabilities* $\tilde{q}_{ij} = x_{ij} / \Sigma_{ij} x_{ij}$ and notionally apply these as a prior probabilities to the T_{ij} flow entropy (1.12). Then solve the problem with this revised objective, constrained by the usual origin, destination and travel cost constraints (1.13) to (1.15). As shown in Roy (1987), an interesting result emerges, with (i) the new modelled base period flows T_{ij}' now corresponding *precisely* with the observed base period flows T_{ij}^0 and (ii) the Lagrange multipliers A_i, B_j and β turning out to be *identical* to those computed by the conventional gravity model (that is, implicitly with *uniform* prior probabilities) used for the original gravity model determination of the modelled base period flows T_{ij}. Moreover, it is proved that $\tilde{q}_{ij}$ is the only prior probability which both keeps the Lagrange multipliers invariant and precisely reproduces the observed pattern of flows. This leads to the following recommendations in cases when the gravity model flows T_{ij} at the base period are not a reliable representation of the observed flows T_{ij}^0 at the same period and where significant transport network or pricing policy changes are expected by the projection period :

(i) Calibrate the doubly-constrained gravity model (1.17)-(1.18) for A_i, B_j and β and compute the modelled flows T_{ij}.

(ii) Compute the base period modelled/observed flow ratios $x_{ij} = (T_{ij}^0 / T_{ij})$ and normalise to probabilities $\tilde{q}_{ij} = x_{ij} / \Sigma_{ij} x_{ij}$.

(iii) If in the projection period, the new origins, destinations and travel costs are depicted as O', D_j' and c_{ij}' respectively, solve the following revised problem :

$$T_{ij}' = \tilde{q}_{ij} \; A_i' O_i' B_j' D_j' \exp - \beta c_{ij}' \tag{1.49}$$

with the new unknown balancing factors A_i' and B_j' given recursively via

$$(A_i')^{-1} = \Sigma_j \, \tilde{q}_{ij} \; B_j' D_j' \exp - \beta c_{ij}' \; ; (B_j')^{-1} = \Sigma_i \, \tilde{q}_{ij} \; \tilde{q}_{ij} \; A' B_j' \exp - \beta c_{ij}' \tag{1.50}$$

This approach, when applied to projection, starts with a model which perfectly reproduces the observed base period observations. Also, in contrast to the Fratar approach, it demonstrates sensitivity to transport changes by retaining the β terms. Note that, in the special case when just the origin and destinations sums change by the projection period but the transport costs remain unchanged, the above combined procedure reduces to the Fratar approach. The author has found the combined method especially suitable for modelling hospital admissions, where accessibility is just one factor considered by doctors in their referrals. Similarly, it has turned out to be most useful in modelling interregional migration, where distance moved is usually important, but not necessarily paramount.

Note that, the prior does not need to be associated with a historical pattern of flows, and an information theory expression such as (1.45) can emerge from an entropy microstate procedure. For instance, in the entropy expression (1.33) for the multi-stop tour model, the 'scaling term' n_{ks} is analogous to a prior probability. In Snickars and Weibull (1977), a separately derived demand model was treated as the prior, with the result being interpreted as a consumption pattern under imposed supply constraints. Although this two-stage procedure may be plausible in some cases, it could be preferable to seek formulations which can be contained within a single stage. This is because, when a new piece of information is added via a constraint, all the Lagrange multipliers associated with any other contemporary information also change. In other words, it may be preferable to replace the information represented in the prior model by constraints, applied simultaneously with the supply constraints. However, more numerical testing should be performed to guide the choice between a one-stage and two-stage procedure.

1.3.2 A Retail Equilibrium Model

From the microstate derivations and the discussion on retail models towards the end of Sect. 1.2.1, a model of retail/service trips S_{ij} between zones i and facilities j can be expressed as

$$S_{ij} = O_i W_j^{\alpha} exp - \beta c_{ij} / \Sigma_j W_j^{\alpha} exp - \beta c_{ij} \tag{1.51}$$

where O_i is the number of trips emanating from zone i, W_j is the floorspace in facility j, β is the gravity impedance parameter, c_{ij} is the generalised cost of travel between zone i and facility j and α is a floorspace scaling parameter calibrated simultaneously with β using the method of maximum likelihood.[4] In a planning context, an important concern is not just customer preferences, but the long term stability of the floorspace location pattern for the operators of the facilities. This is the issue addressed by Harris and Wilson (1978). They postulated that a stable location pattern would occur when the number of customers $\Sigma_i S_{ij}$ visiting each facility j is proportional to the floorspace W_j, that is

$$\Sigma_i S_{ij} = k W_j \qquad \forall j \tag{1.52}$$

where, if there is a total (exogenous) amount of floorspace W to be allocated, the the constant k is given as (S / W), where S is the total number of trips. If (1.51) is substituted for S_{ij} into (1.52), we then have a set of non-linear equation to evaluate the equilibrium floorspace values, say $\hat{W}_j$. From the original paper, it is seen that stable patterns occur when the scaling parameter α is ≤ 1. This implies the absence of scale economies in the perception of floorspace quantities from the point of view of the shoppers. One of the most interesting features of the paper was the identification of *bifurcations* in the approach to equilibrium for certain ranges of α and β. For instance, upon changes of these parameters, one could move from say a more *concentrated* pattern of facilities to a more *dispersed* pattern. The identification of the potential for such changes represents a fundamental contribution to modelling the evolution of urban form.

With over twenty years having elapsed since this key paper appeared, it is

[4] The procedure for calibration of α is summarised in Appendix 1.1.

challenging to try to enhance the approach to fit more generally within the short run and long run equilibrium frameworks in economics. In the short run, the individual retail tenants operating at each centre j can be treated as profit maximizers, either via a simple deterministic model or via a probabilistic entropy model with a profit constraint. Then, a more sophisticated version of (1.51), which includes the effect on demand of price differences in goods and services between the competing facilities, would be developed. Prices would then be gradually adjusted at each facility until the quantities of goods and services supplied at each facility equalled those demanded by the customers. Then, taking a large number of retail tenants at each centre, we postulate that a long run non-cooperative (Nash) locational equilibrium would occur when the profits per unit floorspace are uniform over all facilities, that is, when no individual tenant could unilaterally improve his position by moving to another facility. As pointed out by Kiyoshi Kobayashi[5], if the short run retailer supply problem were probabilistic (eg. derived via entropy) rather than deterministic, the long run equilibrium criterion should denote equal *expected* profits per unit floorspace, rather than equal *average* profits. The enhanced demand relationships would be embedded into these profit relations under an assumed set of prices and we would solve for the corresponding equilibrium floorspace levels. However, with any such set of new floorspace values, we would need to re-adjust prices to re-establish the short run equilibrium. Then, new equilibrium floorspace values would be re-evaluated from the equi-profit relations under the changed prices, and so on, until we have simultaneously established short run price equilibrium for the customers vs retail tenants, as well as long run locational equilibrium for these same tenants. In Chap. 4, a more advanced retail model framework is developed under the above guidelines.

1.3.3 Probabilistic Entropy vs Random Utility for Individual Choice

During the 1970's, considerable controversy arose about the behavioural relevance of so-called 'aggregate models' of the entropy type and the recently developed area of individual choice models based on random utility. As demonstrated in a key article by Anas (1983), this debate was often spurious, and based more on 'historical inertia and intellectual bias', than on a true appreciation of the similarities and differences between the two approaches. Roy and Lesse (1981) showed the structural similarity between classical entropy models and the multinomial logit model when both are estimated on aggregate data. Anas (1983) went further. He used Shannon entropy defined at the individual level to demonstrate not only the structural similarity of the resulting models to those derived from individual choice theory, but also to prove that the parameters of the two models are *identical* when the choice models are estimated by the usual maximum likelihood method. In the following, we adapt the proof of Anas to the 'equivalence' of an entropy model estimated for individual commuters proceeding to job zones, and a random utility model, where choice of job zone is jointly

---

[5] University of Kyoto, Private Communication

defined for an individual commuter at his or her place of residence. So long as each commuter gives his perceived travel cost to his *individual job* in any of the job zones, it is practical to reduce the dimensionality of the problem by performing some aggregation over jobs.

Let $p_{j|h}$ be the *modelled* conditional probability that commuter h, sampled at place of residence, works in zone j and $\delta_{hj} = 1$ if commuter h is *observed* to work in zone j and zero otherwise. Also, let c_{hj} be the perceived cost of travel by commuter h to his individual job in zone j. If there are H households in the sample, $p_h = (1/H)$ is the (exogenous) probability that any household chosen at random is household 'h'. Then, using Shannon entropy, as described in Section 1.2.1, the entropy maximization objective becomes

$$S = \Sigma_h p_h \; \Sigma_j p_{j|h} \,(\log p_{j|h} - 1) \tag{1.53}$$

under the 'origin' constraints

$$\Sigma_j p_{j|h} = 1 \tag{1.54}$$

Destination constraints are applied in the form

$$\Sigma_h \; p_h \, p_{j|h} = \Sigma_h \, p_h \, \delta_{jh} \tag{1.55}$$

Finally, we have the classical transport cost constraints

$$\Sigma_{hj} \, p_h \, p_{j|h} \, c_{hj} = \Sigma_{hj} \, p_h \, \delta_{hj} \; c_{hj} \tag{1.56}$$

If the objective (1.53) is maximized in terms of the flow probabilities $p_{j|h}$ under constraints (1.54) with multipliers λ_h, (1.55) with multipliers η_j and (1.56) under multiplier β, we obtain after eliminating λ_h

$$p_{hj} = exp - (\eta_j + \beta \, c_{hj}) \, / \, \Sigma_j \, exp - (\eta_j + \beta \, c_{hj}) \tag{1.57}$$

which clearly is identical in form to the aggregate classical doubly-constrained gravity model expressed in (1.16). We have replaced the origin zones i of the classical model with individual commuters at their residences. Parameters η_j can be evaluated by the usual iterative technique and β via linear extrapolation.

Following Anas, a 'corresponding' multinomial logit model is developed from random utility theory. Firstly, it is necessary to define a linear stochastic utility function in the form

$$\hat{U}_{hj} = \lambda_h' + \eta_j' + \beta' \, c_{hj} + \varepsilon_{hj} \tag{1.58}$$

where λ_h' is specific to households, η_j' is an alternative-specific constant, β' weights the transport costs and ε_{hj} is the random error term. The latter is taken as independently and identically distributed according to the Gumbel distribution. If the probability of commuter h choosing zone j is taken as the probability that its perceived utility is greater than that of any other alternative, then the utility maximization process yields

$$p_{j|h} = exp - (\eta_j' + \beta' c_{hj}) / \Sigma_j exp - (\eta_j' + \beta' c_{hj}) \tag{1.59}$$

Now if the log likelihood function $\Sigma_{hj}\, p_h\, \delta_{hj}\, log(p_h\, p_{j|h})$ is maximized in terms of the unknown parameters η_j' and β' upon substitution from (1.59), it easy to show that $\eta_j' = \eta_j$ and $\beta' = \beta$. Not only do the respective maximum likelihood equations turn out to be identical to constraints (1.55) and (1.56) of the entropy maximization problem, but (1.59) is also identical in structure to (1.57).

Although Anas has convincingly shown the 'equivalence' of the entropy and utility approaches under the assumptions stated, the author prefers to use a utility approach where data on *individual* consumers is available. On the other hand, when we are restricted to *aggregate* data, such as Census data, or when the sample of individuals is too small to handle problems with multi-dimensioned events, the entropy microstate procedure is more appropriate, having less prior assumptions than a utility approach classically defined for the individual consumer.

1.3.4 Cost Minimization with an Entropy Constraint

In Erlander (1980), the conventional entropy maximization problem is turned on its head. The cost constraint, such as in (1.15), becomes the objective, with the entropy objective, such as in (1.12), being applied as a constraint. We should note that this is not a *dual* formulation in the mathematical sense - it is rather an expression of the interchangeability of objective and constraint in Lagrangian analysis. In Erlander's framework, the minimization of cost is restored, from being a constraint in the entropy model, to being the objective in its deterministic analogue. Then, to modify the solution to pick up the *dispersion* from the deterministic solution, the modelled entropy is induced to reproduce the entropy evaluated at the observed values of the flows. The derivation of Erlander's model is illustrated on the doubly-constrained journey-to-work model, given in (1.12) to (1.18). Let the objective C be the total transport cost to be minimized

$$C = \Sigma_{ij}\, T_{ij}\, c_{ij} \tag{1.60}$$

which occurs in constraint (1.15) in the usual solution. Then, knowing the observed base period flows T_{ij}^0, let the 'observed' entropy S^0 be computed from these flows and applied as the following constraint

$$\Sigma_{ij} T_{ij} [\, log\, T_{ij} - 1\,] = \Sigma_{ij} T_{ij}^0 [\, log\, T_{ij}^0 - 1\,] = S^0 \tag{1.61}$$

which is the objective (1.12) in the conventional case. Now, if the total transport cost C is minimized under the usual origin and destination constraints (1.13) and (1.14) plus the entropy constraint (1.61), we obtain the following

$$T_{ij} = exp - (\lambda_i'' + \eta_j'' + \beta'' c_{ij}) \tag{1.62}$$

where $(1/\beta'')$ is the Lagrange multiplier on the entropy constraint (1.61) and λ_i'' and η_j'' those on the origin and destination constraints (1.13) and (1.14) respectively. This relation can be compared with the conventional solution of (1.16), where the functional form is identical, but the parameters are different. This difference in parameters is an empirical matter - if the model is a relatively good fit to the observed flows, the parameter differences should be small. They would only expected to be identical when the modelled flows precisely reproduce those observed - an almost impossible situation.

With the results from the two alternative procedures being usually quite close to each other and the functional forms being identical, which is to be preferred? This author would prefer the usual microstate procedure, as the entropy maximization objective arises from the principles of statistical mechanics and the constraints from the behavioural foundations of the problem - that is, the key role of transport times and costs in trip distribution. However, this is more a philosophical question which permits of several interpretations. At the same time, and perhaps more importantly, the computation of the 'observed' entropy S^0 requires knowledge of the complete set of observed flows T_{ij}^0. On the other hand, evaluation of the average observed travel cost c^0 may be performed using an appropriate sampling procedure, without needing to know the observed flow distribution T_{ij}^0. In Census years T_{ij}^0 may be available - at other times expensive surveys would be necessary. Thus, data availability may be the main reason for staying with the usual method.

Exercise 1.4 Let there be a 2-zone system, with a retail facility concentrated at the centre of each zone. Assume that the retail model (1.51), taking the floorspace power α as unity, is derived from maximization of the following entropy function: $S = \Sigma_{ij} S_{ij} [log (S_{ij}/W_j) - 1]$, where S_{ij} are the flows and W_j the floorspace at each facility j. The model also needs to satisfy origin constraints $\Sigma_j S_{ij} = O_i$, where O_i are the total trips emanating from zone i. It also satisfies a travel cost constraint $\Sigma_{ij} S_{ij} c_{ij} = S c^0$ associated with a Lagrange multiplier β, where S is the total number of trips, c_{ij} the travel cost between zone i and facility j, with c^0 being the average travel cost per trip as measured at the base period. Assume that: Floorspace: $W_1 = 5000$, $W_2 = 8000$, Trip origins: $O_1 = 40000$, $O_2 = 60000$, Trip costs: $c_{11} = 1.0$, $c_{12} = c_{21} = 2.5$, $c_{22} = 1.5$, Observed flows: $S_{11}^0 = 30000$, $S_{12}^0 = 25000$, $S_{21}^0 = 12002$, $S_{22}^0 = 47998$. Proceed as follows:

(i) From the observed trip flows and transport costs, confirm that the observed average trip cost c^0 is 1.57002.

(ii) Confirm that a gravity parameter β of unity allows the model (1.51) to exactly fit the observed average trip cost c^0, and compute the modelled flows S_{ij}.

(iii) Compute the system entropy S^0 as defined above at the observed flows. Realising that a form such as (1.51) (with $\alpha = 1$) should also arise from the entropy-constrained problem, but with a different parameter β^*,

obtain β^* and the corresponding flows for this problem (*Hint:* Start with a trial value of β^* of unity, corresponding to that for the conventional problem, evaluate the entropy and compare with S^0. If different, adjust β^* iteratively until the modelled flows reproduce S^0). Compare the modelled flows with those obtained in the conventional solution in *(ii)*. Note. With 4 unknown flows, two origin constraints and one behavioural constraint, either on trip cost or entropy, the system has effectively only *one* degree of freedom.

The Erlander analysis above relates to a single event situation, such as destination choice. It remained for Boyce *et al.* (1983) to extend Erlander's ideas for distributions with at least two interdependent events (eg. choice of job and choice of travel mode for the journey to work). They added several entropy constraints according to the number of events. Roy and Lesse (1985), inspired by Boyce's work, produced a more orthodox entropy maximization variation of the same models, at the same time recognising that the models possessed a similar structure to nested logit models arising from random utility theory. These points are illustrated below on a simple housing choice model.

The following analysis is extracted, with some minor modifications, from Roy and Lesse (1985). Consider the process of choice of housing in zones i by workers with jobs in zones j. We assume that the spatial distribution of jobs is given exogenously, with $\bar{p}_j$ being the probability that any worker chosen at random has a job in zone j. Let p_i be the unknown probability that a worker resides in zone i and $p_{j|i}$ be the conditional probability that given that a worker chooses a house in zone i he or she works in zone j, with $p_{ij} = p_i \, p_{j|i}$ being the corresponding joint probability. It is also assumed that the workers are segmented into income groups (ideally they should also be allocated to households). Finally, for any such income group, let the average housing rent paid in each zone i be denoted by r_i, the average generalised journey-to-work costs by c_{ij}, with $\bar{r}$ being the average sum of rent and transport costs paid by any member of the given income group, evaluated from base period data, which the estimated model must replicate. Note that, the introduction of such a *constraint* on the sum of rent plus commuting costs into our probabilistic model yields a probabilistic form of the deterministic model of Herbert and Stevens (1960), where the minimization of $\bar{r}$ is the *objective.*

Decomposing the entropy of the joint probability distribution p_{ij} into its two component parts, the Shannon entropy *maximization* objective is given as

$$S = -\Sigma_i p_i \Sigma_j p_{j|i} (\log p_{j|i} - 1) - \Sigma_i p_i \log p_i \qquad (1.63)$$

Note that, the unity term, which sums to a constant, has been omitted from the second entropy expression for reasons of elegance. The above entropy is to be maximized jointly in terms of p_i and $p_{j|i}$. The first constraint, with Lagrange multiplier Ω is the probability normalisation condition

$$\Sigma_i p_i = 1 \qquad (1.64)$$

Analogous constraints, with multipliers ω apply for the conditional probabilities

$$\Sigma_j \, p_{j|i} = 1 \qquad \forall i \tag{1.65}$$

The worker probabilities $\bar{p}_j$ are reproduced via

$$\Sigma_i \, p_i \, p_{j|i} = \bar{p}_j \qquad \forall j \tag{1.66}$$

Then the key behavioural constraints, with multiplier α are introduced

$$\Sigma_{ij} \, p_i \, p_{j|i} \, (r_i + c_{ij}) = \bar{r} \tag{1.67}$$

Up until now, the above model represents a conventional bi-variate Shannon entropy model. Now we consider, what if the level of uncertainty of the p_i entropy is considerably different from that of the $p_{j|i}$ entropy? Surely, we should then be maximizing a *weighted* sum of the component entropies, rather than the pure sum in (1.63)? The 'weighting' is *endogenously* determined by introducing the following constraint on the base period entropy of p_i

$$\Sigma_i \, p_i(\log p_i - 1) = \Sigma_i \, p_i^0 (\log p_i^0 - 1) \tag{1.68}$$

where p_i^0 represent the observed housing choice probabilities at the base period. For cosmetic reasons (to illustrate the analogy with nested logit), a Lagrange multiplier of $-(1-\theta)/\theta$ is associated with constraint (1.68).

The first step is the maximization of (1.63) in terms of $p_{j|i}$, absorbing the p_i into the multipliers ω_i, which are eliminated to yield

$$p_{j|i} = \exp - [\eta_j + \alpha(r_i + c_{ij})] \, / \, \{\Sigma_j \exp - [\eta_j + \alpha(r_i + c_{ij})]\} \tag{1.69}$$

Upon differentiation of (1.63) with respect to p_i, substitution of (1.69) for $p_{j|i}$ and after considerable cancellation, we obtain the result

$$p_i = [\Sigma_j \exp - [\eta_j + \alpha(r_i + c_{ij})]^\theta \, / \, \{\Sigma_i \, [\Sigma_j \exp - [\eta_j + \alpha(r_i + c_{ij})]^\theta\} \tag{1.70}$$

Upon recognition of the numerator of (1.70) as some sort of 'composite' average over j of the rents, job attractiveness and transport costs associated with each housing zone i, it is possible to define 'inclusive value' rents r_i' in the form

$$r_i' = -(1/\alpha) \log \Sigma_j \, [\eta_j + \alpha(r_i + c_{ij})] \tag{1.71}$$

enabling (1.70) to be given as

$$p_i = \exp - \theta\alpha \, r_i' \; (\Sigma_i \exp - \theta\alpha \, r_i') \tag{1.72}$$

The structure of (1.69) and (1.72) clearly is analogous to that obtained via a two-level nested logit model (Hensher and Johnson, 1980), with θ the inclusive value parameter. The differences are that it is derived via entropy maximization and is estimated on aggregate data, albeit with some market segmentation. This derivation can be seen to be an enhancement of the work of Anas (1983), where entropy models and the simpler multinomial logit models were shown to share a similar structure.

1.3.5 Some Key Conceptual Contributions

In this section, we want to briefly summarise some of the major contributions to spatial interaction modelling made by two key researchers, Tony E. Smith and Stewart A. Fotheringham, as well as some related work by the author.

Some Contributions by Tony E. Smith. One of Smith's first major efforts was the development of a cost-efficiency principle for spatial interaction (Smith, 1978). In this highly formal paper, Smith aimed to establish a more *behavioural* and intuitively plausible basis for spatial interaction modelling. The hypothesis is as follows: "A trip distribution P on X is said to satisfy the *cost-efficiency principle* if and only if for each possible (travel) cost configuration c on X and pair of trip patterns consistent with the same trip activity A on X, the pattern with the lowest travel cost is always the most probable one." Based on this definition of the principle, an *Equivalence Theorem* is proved: "A trip distribution P satisfies the cost-efficiency principle if and only if P is representable by an exponential gravity model." A corollary is that "All activity-equivalent trip patterns with the same total travel cost must be equally probable." Thus, a behavioural basis is provided for the exponential gravity model, independent of that implied by the conventional entropy maximization technique, but consistent with its results.

In Smith (1989), the general interaction model of Alonso (1978) was examined. An important result was that the form of Alonso's model where the total number of interactions is exogenous, as also discussed by Fotheringham and Dignan (1984), can be simplified to the following

$$T_{ij} = g\, A_i^{\alpha\text{-}1}\, B_j^{\beta\text{-}1}\, f_{ij} \tag{1.73}$$

where T_{ij} is the number of trips between i and j, g is a normalisation parameter, A_i and B_j are sets of *systemic variables,* interpreted as accessibilities, defined at the origins and destinations respectively, α and β are positive scalars lying between zero and one and the f_{ij} are exogenous linkage factors between i and j. The Alonso model is shown to be interpretable within the decision theory developed in the same paper. The current author is still uncertain about the relative merits of Alonso's model in relation to other models of spatial interaction. Certainly, Smith's arguments support the 'reduced form' of (1.73), which resembles several models discussed earlier in this chapter.

Although Smith was a co-author of the book Sen and Smith (1995), as mentioned earlier, his most recent major *personal* contribution of relevance to this book was the development of the MPS (Most Probable State) approach in Smith (1990). "The fundamental hypothesis of MPS-analysis is that all micro population behaviour is describable probabilistically in terms of certain of its macro properties." In Smith's terminology, a *probabilistic theory of micro behaviour* (that is, the *'theory'*) is defined as a function of the macro properties (eg. total travel cost), which are themselves given as functions of the trip variables. The chosen macro properties are not selected at random, but 'are inextricably bound up with the *theory* itself'. The important results, proved rigorously by Smith, are stated qualitatively here, as follows:

If theory T is true, then we can compute an exact form for the conditional probabilities of macro states given any observed values of the macro properties.

The conditionally most-probable macro states are precisely those which can occur in the largest number of ways (as in the entropy microstate approach).

If theory T is true, then for all sufficiently large population realisations, the maximum entropy state is the ***overwhelmingly most-probable macro state*** *which can occur.*

The last statement represents a major generalisation of the restricted proof in Wilson (1967), which was just demonstrated for an entropy model with only one macro property, that is, knowledge of the total population in the system. This generalisation is fundamental to the statistical validity of the entire approach. Note also that, testing of any MPS-based probabilistic theory does **not** require that the samples be statistically independent.

To continue, 'Each choice of macro variables constitutes a *different* theory of micro behaviour. For instance, total kinetic energy is defined as a macro property in Boltzmann's probabilistic equilibrium theory of an ideal gas, lying at the very heart of the theory.'

This author has found Smith's work of paramount importance in attempts to generalise spatial interaction models to include microeconomics concepts. For instance, in developing a probabilistic model for the supply behaviour of profit-maximizing competitive firms, profit is the key macro property. In fact, in Chap. 3, we somewhat generalise Smith's results for the case where the basic 'theory' is deterministic (i.e. the theory of the competitive firm), but where we want to develop a more general probabilistic version. The key results from Evans (1973) are adapted to indicate that the results of any classical deterministic theory are obtainable as limiting cases of the probabilistic model in which the relations of the deterministic theory are introduced as a macro constraint reflecting observations. This implies, for instance, that we can use entropy or MPS analysis to produce probabilistic versions of many of the deterministic models of economics. Moreover, as the entropy (uncertainty) approaches zero, the solution of the probabilistic model converges asymptotically towards that of its deterministic counterpart. Thus, the work of Smith and Evans is pivotal in achieving our key aim - introducing microeconomics concepts into spatial interaction modelling.

Some Contributions of A. Stewart Fotheringham. Commencing in the early 80's, Fotheringham in Fotheringham (1983a, 1983b, 1986) and Fotheringham and

Knudsen (1986) made some far reaching contributions to spatial interaction modelling. One of his major theses is that spatial choices are not independent of spatial 'structure', that is, the pattern of spatial opportunities in a given area. If this is true, then some gravity models, such as the origin-constrained model used for retailing studies, which have an independence structure, are mis-specified. For instance, as discussed in Fotheringham (1983b), when a system with say two identical retail centres spaced well apart is enhanced by the addition of an identical third retail centre, will the *relative* competitive position of the two existing retailers be influenced by alternative location options of the third centre?

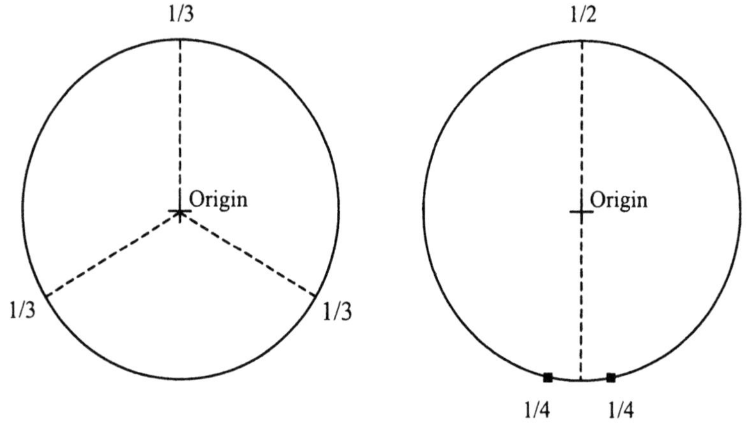

Fig. 1.2 Alternative Retail Centre Configurations

The conventional model would say no! However, what if the third retail centre were located quite close to say centre number *1*? As illustrated in Fig. 1.2, if each centre were equally accessible to customers, then the conventional model would indicate that addition of the third centre would change the market shares from *{1/2, 1/2}* to *{1/3, 1/3, 1/3}*, with no effect of the proximity of centres *1* and *3* on the relative competitive position of the original two centres. Intuitively, for this case of identical centres of equal accessibility, one may expect the market share of centre *1* to be split with that of the new centre, leading to shares of *{1/4, 1/4, 1/2}*! The appreciation of this anomaly, known as Independence from Irrelevant Alternatives (IIA), gave Fotheringham the incentive to develop new forms of spatial interaction models which are in some sense hierarchical, as given explicitly in Fotheringham (1986), whereby customers are not only attracted to a particular centre per se, but to the *milieu* of that centre with respect to adjacent centres. In this way, both spatial competition and spatial agglomeration influences can be identified. It should also be noted at this point, that inspired by Fotheringham's 1983 work, Roy (1985) developed via entropy a hierarchical destination choice or 'cluster' model, which shares the non-IIA property of the *nested logit* random utility model and overcomes the above anomaly. In the approach of Roy (1985), there is a pre-determined nested hierarchical choice between a cluster and then a particular destination within that cluster, with the order of choice not being obtained statistically, as in nested logit models. Fotheringham's formulation, on the other hand, influences the choice of a primary destination by the accessibility of a set of alternative secondary destinations to this primary destination.

In illustrating Fotheringham's ideas, we turn to his new competing destinations retail model presented, for instance, in Fotheringham and Knudsen (1986). In contrast to the conventional retail model in (1.51), upon defining flows S_{ij} in terms of trip origins O_i, floorspace W_j and home-based travel costs c_{ij}, the competing destinations model is expressed as

$$S_{ij} = O_i W_j^{\alpha} A_j^{\gamma} exp - \beta c_{ij} / [\Sigma_j W_j^{\alpha} A_j^{\gamma} exp - \beta c_{ij}] \tag{1.74}$$

with the 'potential measure A_j defined for competing destinations as

$$A_j = \sum_{l \neq j} W_l c_{jl}^{-\beta} d_{jl} \tag{1.75}$$

where c_{jl} represents the unit travel cost between the *primary destination j* and a *nearby destination l*. The binary variables d_{jl} are unity when destination l lies within a pre-specified range of destination j and zero otherwise. The introduction of the 'competing destinations' potential terms A_j^{γ} distinguishes the new model from the conventional model in (1.51). A positive value of the scaling index γ demonstrates the presence of consumer agglomeration economies, with a negative value indicating the presence of consumer competition or congestion forces.

The above model structure clearly represents a great step forward in recognition of interdependencies in spatial choice. It is also supported by a great deal of empirical work by Fotheringham and others. At the same time, there are a couple of 'cosmetic' issues which deserve attention, such as (i) the absence of a floorspace scaling power α on the floorspace W_l of the competing destinations l and (ii) the use of the power form for the impedance of the competing destinations, rather than the exponential form used for the primary destinations j. Some advances in expanding the competing destinations model into the framework of multi-stop activity analysis, certainly *implied* by the structure of Fotheringham's models, were made by Thill (1995). Some further progress is described below.

An Activity-Based Competing Destinations Framework: Although recent work of the author will generally be covered from Chap. 3 onwards, a microstate-based competing destinations activity model has loomed as a challenge for some years, starting with Roy (1985). In other words, we would like spatially interdependent model structures, such as those of Fotheringham, to emerge as the *results* of appropriate maximization problems, as an alternative to postulating appropriate functional forms. These formulations are distinguished by the nature of the choice process for the secondary destination(s) given that a primary destination is chosen, described as follows.

Just One Secondary Destination per Trip: Consider that at each primary destination j, there are n_j destinations k which may be chosen. If the number of trips which proceed to a secondary destination k via primary destination j is S_{ijk}, then the total number of trips S_{ij} which stop at j is simply given as $(\Sigma_k S_{ijk})$. The number of microstates Z is given as

$$Z = \pi_i \{O_i! / [\pi_{jk} S_{ijk}!]\} \tag{1.76}$$

This yields the variable terms of the entropy S in the usual way

$$S = -\Sigma_{ijk} S_{ijk} [\log (S_{ijk}/W_j^\alpha W_k^\alpha) - 1] \tag{1.77}$$

where the scale effects of floorspace now include both the visited centres j and k (assuming the same index α. for each). The origin constraints are now given as

$$\Sigma_{jk} S_{ijk} = O_i \tag{1.78}$$

As the trips are assumed to proceed from i to j and on to k, returning the same way, the travel cost constraint is given as

$$\Sigma_{ijk} S_{ijk} (c_{ij} + c_{jk}) = S\,c \tag{1.79}$$

where c is the observed average cost of each combined trip. Upon maximization of S in terms of the S_{ijk}, under the given constraints, we obtain

$$S_{ijk} = O_i W_j^\alpha W_k^\alpha \exp - \beta (c_{ij} + c_{jk}) / [\Sigma_{jl} W_j^\alpha W_l^\alpha \exp - \beta (c_{ij} + c_{jl})] \tag{1.80}$$

Upon setting the potentials A_j as the summations

$$A_j = \Sigma_l W_l^\alpha \exp - \beta c_{jl} \tag{1.81}$$

further summation of S_{ijk} over $k = 1, n_j$ yields

$$S_{ij} = O_i W_j^\alpha A_j \exp - \beta c_{ij} / [\Sigma_j W_j^\alpha A_j \exp - \beta c_{ij}] \tag{1.82}$$

There are remarkably small differences compared with Fotheringham's model in (1.74) and (1.75), including (i) absence of the power γ on the potential, (ii) the use of limited summations $k = 1, n_j$ in defining A_j, rather than use of the binary variables d_{jl} to define the summation and (iii) inclusion of the floorspace index α within the definition of A_j. Whereas point (ii) is 'cosmetic', it does seem quite plausible to have a floorspace scale effect other than unity on the secondary destinations l. This could well be different from the index α chosen for the primary destinations, and could be calibrated via maximum likelihood. On the other hand, it does seem reasonable to include Fotheringham's agglomeration/

competition index γ on the potential terms, recognising that the sum $\Sigma_k S_{ijk}$ of the trips to all secondary destinations may be less than or greater than the number of primary trips S_{ij}. But, as seen in the next section, the recognition of primary trips involving the choice of several secondary destinations (or none at all) yields a quite different 'potential' function to that chosen by Fotheringham.

Choice Set Includes Potentially All Secondary Destinations Consider that the shoppers decide whether or not to proceed to each secondary destination in turn. In this case $S_{ij} \neq \Sigma_k S_{ijk}$ and the microstate definitions must reflect this, and the S_{ijk} flows now represent those of the S_{ij} primary flows which decide to proceed to secondary destination k. Consider, as before, that the O_i trips generated from zone i choose alternative centres j in groups S_{ij}. Then, each such primary trip is taken to have the option of choosing in turn between zero and the total number n_j of the secondary destinations k, with S_{ijk} shoppers choosing k and $(S_{ij} - S_{ijk})$ not choosing k. Thus the microstates Z reflect these potential multiple choices and are given as

$$Z = \pi_i \{O_i! / [\pi_j S_{ij}!]\} \{\pi_{ijk} S_{ij}! / [S_{ijk}! (S_{ij} - S_{ijk})!]\} \tag{1.83}$$

Adopting the usual approximations as well as the W attractiveness terms in terms of floorspace, the entropy comes out as

$$S = - \Sigma_{ij} S_{ij} [\log (S_{ij}/W_j^{\alpha} - 1)] + \Sigma_{ij} n_j S_{ij} [\log S_{ij} - 1] - \Sigma_{ijk} S_{ijk} [\log (S_{ijk}/W_k^{\alpha} - 1)] \\ - \Sigma_{ijk} (S_{ij} - S_{ijk}) [\log (S_{ij} - S_{ijk}) - 1] \tag{1.84}$$

where n_j represents the potential for multiple choices of secondary destinations and the secondary trips $(S_{ij} - S_{ijk})$ not taken are not associated with any attractiveness term. Noting that the trips S_{ij} and S_{ijk} are not here defined in terms of each other, the objective (1.84) must be maximized simultaneously in terms of both the primary flows S_{ij} and the secondary flows S_{ijk}. The origin constraints are here

$$\Sigma_j S_{ij} = O_i \tag{1.85}$$

It is also useful to provide the calibration process with data on the observed number of trips S^*_{jk} proceeding from centre j to each secondary destination k - these constraints will be released in the projection period. Thus, we have

$$\Sigma_i S_{ijk} = S^*_{jk} \tag{1.86}$$

Finally, with c^* being the observed cost per multi-stop trip, returning and re-starting always via the primary destination, the travel cost constraint is given as

$$\Sigma_{ij} S_{ij} c_{ij} + \Sigma_{ijk} S_{ijk} c_{jk} = c^* S \tag{1.87}$$

With η_{jk} taken as the Lagrange multiplier on (1.87), the maximization process yields after some manipulation

$$S_{ijk} = S_{ij} W_k^{\alpha} exp - (\eta_{jk} + \beta c_{jk}) / [1 + W_k^{\alpha} exp - (\eta_{jk} + \beta c_{jk})] \qquad (1.88)$$

and for the primary flows

$$S_{ij} = O_i W_j^{\alpha} A^*_j exp - \beta c_{ij} / [\Sigma_j W_j^{\alpha} A^*_j exp - \beta c_{ij}] \qquad (1.89)$$

in which the potentials A^*_j now assume **the product form**

$$A^*_j = \pi_k [1 + exp - (\eta_{jk} + \beta c_{kj})] \qquad (1.90)$$

with the product k going over all secondary destinations $k = 1, n_j$ available to centre j. Although (1.89) is similar in 'form' to (1.82) derived for just one secondary stop, the definitions of the potentials A_j in (1.81) and (1.90) are quite different. The product form in (1.90) is seen to arise from the 'synergy' of being able to make *multiple* stops at secondary destinations via the *one* primary destination. As this possibility clearly exists in general, the above more general form may well prove more useful in practice than the original Fotheringham model. At the same time, the choice set *perceived* by each shopper may not cover *all* the secondary destinations k chosen as 'accessible' to the primary destination j, as assumed in (1.90) above. In order to account for this, an unknown agglomeration/competition index may be introduced on our product potential $\overline{A}_j$, analogously as performed by Fotheringham for his summation potential A_j. This index would be calibrated, as for α, via a simultaneous maximum likelihood step. The existence of agglomeration vs competition effects may also be captured by use of a two-level 'nested entropy' approach (Roy, 1985), where the model can endogenously 'weight' the entropy (uncertainty) of choice of a primary destination vis-à-vis the entropy of choice of secondary destinations conditional on choice of the primary destination. The above derivation provides a good demonstration of the versatility of our entropy microstate procedure, now interpreted with more rigour within the MPS approach of Smith (1990).

The reader will perceive that the above models all relate to demand. However, multi-stop and multi-purpose trips clearly have implications for retail centre managers in their choice of the mix of goods and services to be provided at centres under their control. In Chap. 4, a companion supply-side model is developed in an attempt to come to terms with these practical issues.

1.3.6 Characterisation of Model Errors and Robustness

In examination of the potential of probabilistic models for projection, four main criteria may be identified: (i) goodness-of-fit of the estimated model with respect to the observations, (ii) some measures of how the data and their use in the model reduces the level of uncertainty of the modelled distribution compared with that from the chosen null hypothesis, yielding measures of robustness, (iii) a key asymptotic property and (iv) use of diagnostic statistics from maximum likelihood. These criteria are now treated in order.

Goodness-of-Fit Measures. Experience with calibration of spatial interaction models has convinced the author that the largest *relative* errors with respect to observations occur for the smaller flows. In fact, the *absolute* value of the error is often not much less for the smaller flows than for the larger flows. In many policy contexts, we may be primarily interested to use an error measure which doesn't allow the errors in the smaller flows to dominate. An appropriate measure in this context is the Root Mean Square (RMS) error. If $\boldsymbol{T}$ is the matrix of modelled flows and $\boldsymbol{T}^0$ that of the corresponding observations, then the RMS measure is given as

$$\text{RMS}\,(\boldsymbol{T}) = \{(1/mn)\, \Sigma_m \Sigma_n\, [T_{ij}^{\,0} - T_{ij}]^2\, \}^{1/2} \tag{1.91}$$

where m and n are the number of rows and columns respectively in matrix $\boldsymbol{T}$.

In some cases, we may be more interested in a measure recognising *relative* errors in the model. Then, we may examine a measure ARAE (Average of Relative Absolute Errors) as follows

$$\text{ARAE}\,(\boldsymbol{T}) = \{(1/mn)\, \Sigma_m \Sigma_n\, |T_{ij}^{\,0} - T_{ij}| \,/\, T_{ij}^{\,0}\, \} \tag{1.92}$$

From Snickars and Weibull (1977), it is seen that the more general information models usually have better goodness-of-fit properties than the straight entropy models. This is because they are structured so as to fit the observed distribution 'perfectly' at the base period. So long as behaviour does not change fundamentally by the projection time period, they can potentially produce an excellent fit to observations. Behavioural changes can be accommodated by the enhanced information theory approach of Roy (1987), which nevertheless retains the goodness-of-fit properties of the usual approach.

Measures of Model Robustness. As stated earlier, the introduction of relevant 'information' via constraints to an entropy model reduces the uncertainty of the modelled distribution compared with that from a chosen null hypothesis (eg. the uniform distribution). An index $\boldsymbol{R}$ of robustness may be defined as the relative reduction of uncertainty (entropy), measured from the maximum entropy uniform distribution, caused by the introduction of the full set of constraints to the model, defined as

$$\boldsymbol{R} = [S_{max} - S] \,/\, S_{max} \tag{1.93}$$

There are also correlation measures, measuring the level of correlation between single events and joint events. For example, in Roy (1985), a relative degree of correlation measure $\boldsymbol{C}$ is defined to indicate the information 'transmitted' between events. For instance, in a typical origin/destination model, let S_i be the entropy of the origin distribution, S_j the entropy of the destination distribution and S_{ij} the entropy of the (joint) flow distribution. Then C is defined as

$$C = [S_i + S_j - S_{ij}] / S_{ij}^{max} \tag{1.94}$$

where S_{ij}^{max} is the entropy of the uniform flow distribution. This measure describes the relative influence of transport costs on flow patterns, after accounting for the origin 'push' and destination 'pull' effects.

Key Asymptotic Property. As discussed in Sect. 1.3.5, Smith (1990) in Theorem 5.2 demonstrated that "If the underlying theory T '(e.g. microeconomic theory)' is true, then for all sufficiently large populations, the maximum entropy state is the *overwhelming most probable* macro state which can occur." This yields the great advantage of being able to interpret our solution as the *expected* result.

Diagnostic Statistics from Maximum Likelihood. Baxter (1985) used the correspondence between entropy models and maximum likelihood models (Karlqvist and Marksjö, 1971) to suggest some diagnostic statistics, especially in cases where the Poisson distribution of flows implied by the latter approach is not plausible. Thus, we can use the entropy approach, with its precise consideration of the statistics of the given microlevel and its appropriate behavioural and flow constraints, to yield the most relevant functional form for the flows. Knowing that if this functional form is estimated by maximum likelihood the resulting equations will correspond precisely with the constraints on the entropy problem, we can then avail ourselves of the whole diagnostic armoury associated with the method of maximum likelihood to assess our models.

Solutions to Numerical Exercises

Exercise 1.1

(i) $S_{11} = 5421.7,\ S_{12} = 578.3,\ S_{21} = 475.8,\ S_{22} = 3524.2,\ S_1 = 5897.5,\ S_2 = 4102.5$

(ii) $S_{11} = 5384,\ S_{12} = 616,\ S_{21} = 447,\ S_{22} = 3553,\ S_1 = W_1 = 5831,\ S_2 = W_2 = 4169$

Exercise 1.2

(i) $k = 4,\ s = 3$, yielding $[3!/(2!\ 1!)]^2 = 3^2 = 9$ microstates given as :

$a \rightarrow (1,2)\ b \rightarrow (1,2) \quad a \rightarrow (1,2)\ b \rightarrow (1,3) \quad a \rightarrow (1,2)\ b \rightarrow (2,3)$

$a \rightarrow (1,3)\ b \rightarrow (1,2) \quad a \rightarrow (1,3)\ b \rightarrow (1,3) \quad a \rightarrow (1,3)\ b \rightarrow (2,3)$

$a \rightarrow (2,3)\ b \rightarrow (1,2) \quad a \rightarrow (2,3)\ b \rightarrow (1,3) \quad a \rightarrow (2,3)\ b \rightarrow (2,3)$

(ii) With twice as many *distinguishable* stops being defined compared with above, the number of microstates becomes $\{(2k-2)! \ / \ [(s-1)! \ (2k-s-1)!]\}^{M_k^s}$. For the above example, this yields $\{(6!) / [(2!) (4!)]\}^2 = 225$ microstates

Exercise 1.3

(i) $Z = 4! / \{a! \ b!\}$ Note: $0! = 1$. Values of Z for (a,b) :
$(0,4) \rightarrow 1$; $(1,3) \rightarrow 4$; $(2,2) \rightarrow 6$; $(3,1) \rightarrow 4$; $(4,0) \rightarrow 1$
Thus, $(2,2)$ (that is, $a=2, b=2$) is most probable macrostate.

(ii) $Z = \{2! / [a! \ (2-a)!]\} \{4! / [b! \ (4-b)!]\}$ Values of Z for (a,b) :
$(0,4) \rightarrow 1$; $(1,3) \rightarrow 8$; $(2,2) \rightarrow 6$ (the rest not permissible)
Thus $(1,3)$ is most probable macrostate.

(iii) $Z = \{(2+a-1)! / [(2-1)! \ a!]\} \{(4+b-1)! / [(4-1)! \ b!]\}$ Values of Z for (a,b) :
$(0,4) \rightarrow 35$; $(1,3) \rightarrow 40$; $(2,2) \rightarrow 30$; $(3,1) \rightarrow 16$; $(4,0) \rightarrow 5$
Thus, $(1,3)$ is the most probable macrostate.

(iv) $Z = \{4! / [a! \ b!]\} \{2! \ / (2-a)!\} \{4! / (4-b)!\}$ Values of Z for (a,b) :
$(0,4) \rightarrow 24$; $(1,3) \rightarrow 192$; $(2,2) \rightarrow (144)$
Thus, $(1,3)$ is the most probable macrostate.

(v) $Z = \{4! / [a! \ b!]\} \{2^a\} \{4^b\}$ Values of Z for (a,b) :
$(0,4) \rightarrow 256$; $(1,3) \rightarrow 512$; $(2,2) \rightarrow 384$; $(3,1) \rightarrow 128$; $(4,0) \rightarrow 16$
Thus, $(1,3)$ is the most probable macrostate.

Note. For typical large problems, the maximum entropy distribution would be *different* for all the cases *(i)* to *(v)*.

Exercise 1.4

(i) This is simple arithmetic.

(ii) From (1.51), we have : $S_{11} = 29476$; $S_{12} = 10524$; $S_{21} = 11216$; $S_{22} = 48784$

(iii) $S^0 = -513009.4$. For $\beta^* = 1.002$, modelled entropy $S = -513012$ *(OK)*. The modelled flows are: $S_{11} = 29500$; $S_{12} = 10500$; $S_{21} = 11198.4$; $S_{22} = 48801.6$.
Note. For a real problem, the differences between β and β^* may well be greater.

Appendix 1.1

In Sect. 1.3.2 (see footnote 4), we indicated that the maximum likelihood procedure for estimating the floorspace power α in the Wilson retail model in (1.51) would be provided. As all the unknown parameters are to be determined simultaneously, we can re-state (1.51) in terms of all the unknown Lagrange multipliers in the form

$$S_{ij} = A_i \, W_j^{\alpha} \, exp - \beta c_{ij}$$

where A_i represents the origin 'attractiveness' to be evaluated to satisfy the origin constraints. The maximum likelihood criterion Z is to be minimized in terms of α in relation to observed shopper flows $S_{ij}^{\,0}$ and the modelled flows S_{ij} given by the above model in the form

$$Z = \Sigma_{ij} \, S_{ij}^{\,0} \, log \, [S_{ij}/(\Sigma_{ij} \, S_{ij})]$$

Noting that $d \, W_j^{\alpha} / d\alpha = log \, W_j \, (W_j^{\alpha})$, the extremum equation for α becomes

$$\Sigma_{ij} \, S_{ij}^{\,0} \, log \, W_j - \{(\Sigma_{ij} \, S_{ij}^{\,0})/(\Sigma_{ij} \, S_{ij})\} \, [\Sigma_{ij} \, A_i \, W_j^{\alpha} \, exp - \beta c_{ij} \, log \, W_j] = 0$$

Upon noting that $(\Sigma_{ij} \, S_{ij}) = (\Sigma_{ij} \, S_{ij}^{\,0})$ via the initial conditions, the above relation collapses into the convenient form

$$\Sigma_{ij} \, S_{ij} \, log \, W_j = \Sigma_j \, S_j^{\,0} \, log \, W_j$$

where we only need for calibration of α the observed number of shoppers $S_j^{\,0}$ patronising each centre j rather than the complete set of flows $S_{ij}^{\,0}$ distinguished by origin. The above 'constraint' (after substitution of S_{ij} from above) is then solved for α simultaneously with the usual travel cost constraint for β. As we have good starting values for β and we know that α is rather close to unity, we can usually adopt the linear extrapolation approach for calibration, rather than the more complicated Newton-Raphson procedure.

2 Key Insights in 'Space' and Microeconomics

2.1 Some Important Theoretical Advances

2.1.1 Urban/Rural Model and Urban Land Models

The classical land use models provide an excellent introduction to the role of space in microeconomics, as well as to the differences between short run price adjustments and long run locational equilibria. Two of the most significant models in this class are introduced below.

The Urban/Rural Model of von Thünen. This model was first published in 1826 and further clarified by Launhardt (1885). The reader is referred to von Thünen (1966) for a recent translation. The main purpose of the model is to explain the distribution of land use for different agricultural commodities around a single town, accounting for relative production costs, relative yield per unit area and relative transport costs. The model makes the following simplifying assumptions: (i) the town is a point on a homogeneous plain, (ii) all commodities are shipped to the town, (iii) the transport network is homogeneous with transport costs linearly proportional to distance, (iv) producers pay the transport costs, (v) there are no scale economies in production, (vi) there are two separate crops, (vii) the geometry of the land use is radial, (viii) at any given radius out from the town, only the higher rent-producing commodity is produced and (ix) demand for each commodity is a non-increasing function of price (the classical assumption). Despite the above restrictions, the model is of fundamental importance.

We start with the following notation :

r = Distance out from town

p_{rj} = Equilibrium land rent per unit area of crop j at radius r

Q_j = Yield of crop j per unit area

v_j = Unit selling price for crop j

z_j = Unit production cost for crop j

t_j = Unit transport cost for crop j per unit of distance

r' = Radial extent of inner crop $(j=1)$ if two crops present

r_m = Radial extent of outer crop $(j=2)$ if present (where rent drops to zero)

$D(v_j)$ = Demand function for crop j in terms of its price v_j

Consider first the *long run* locational equilibrium, where profits must be zero at all points. [If profits were positive anywhere, existing producers may shift location or new producers may enter the market. If they were negative, some producers must quit the market.] Thus, at any radius r, the zero profit condition for either crop j implies that the land rent must equal revenue minus production and transport costs, yielding

$$p_{rj} = Q_j (v_j - z_j - rt_j) \tag{2.1}$$

In examining the function (2.1) in terms of the geometry of the system and the question of whether one or two crops exist at equilibrium, refer to Fig. 2.1 below, existing at any radial cross-section.

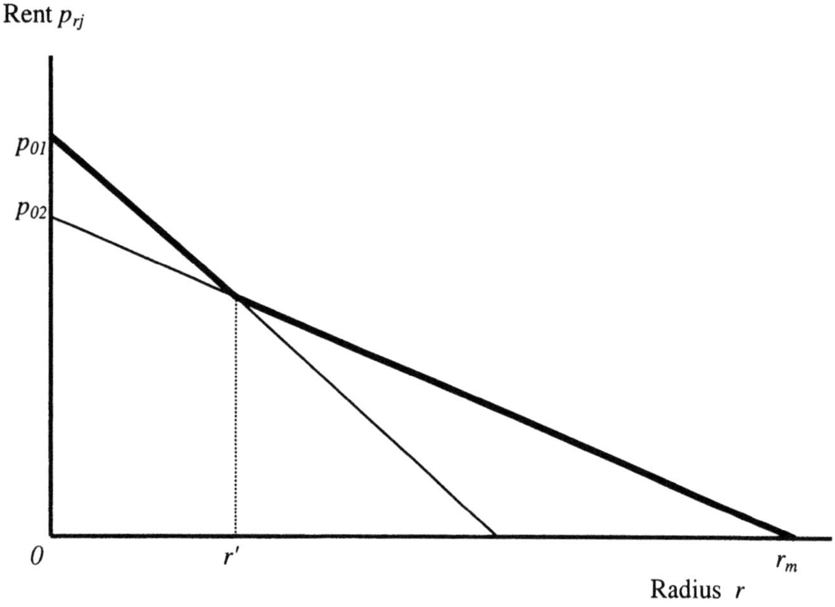

Fig. 2.1 Rent distribution and extent of cultivation over rural hinterland

It is seen from both the function (2.1) and the above figure that two crops can exist in equilibrium if and only if the crop with the greater rent p_{01} at the centre of the circle also has the greater unit transport cost $(Q_1\ t_1)$, that is, the greater rent gradient. Otherwise, the rent lines for the two crops will never intersect and crop 1 alone will be planted. As the rents are endogenous in the overall analysis and the unit transport costs $(Q_j\ t_j)$ are exogenous, the crop with the greater unit transport cost must be denoted as crop *1* in the analysis. Assuming for the moment that two crops may be planted, crop *1* is planted out to radius r' and crop 2 between radius r' and r_m where the rent drops to zero. From Fig. 2.1 and the rent function (2.1), it is seen that the 'intersection' radius r' and end radius r_m come out as

$$r' = (r_{01} - r_{02}) / (Q_1 t_1 - Q_2 t_2) \quad ; \quad r_m = (v_2 - z_2) / t_2 \tag{2.2}$$

where the rents r_{0j} at zero radius emerge from (2.1) as

$$r_{0j} = Q_j (v_j - z_j) \tag{2.3}$$

corresponding to unit revenue minus production costs.

In the *short run*, we seek the prices v_j which ensure that demand equals supply for each crop. As this short run result is to be 'embedded' into the long run conditions (2.1), implying the *simultaneous* existence of both short and long run equilibrium, relations (2.2) and (2.3) must be satisfied for the equilibrium prices. Assuming that demand functions D_j are provided exogenously, the prices must be adjusted such that for crops *1* and 2, demand equals supply S_j in the form

$$D_1 = S_1 = \pi Q_1 r'^2 \quad ; \quad D_2 = S_2 = \pi Q_2 (r_m^2 - r'^2) \tag{2.4}$$

where the latter relation depicts the *annular* distribution of the outer crop *2*. The solution involves the simultaneous solution of (2.2) to (2.4), with crop *1* dropping out if at any stage the rent $r_{02} \geq r_{01}$. Even if the demand functions are linear in price, the non-linearity in (2.4) implies that the overall solution should be carried out iteratively.

What does this classical model demonstrate? First, the commodities with the higher unit (revenue minus production costs) p_{0j} per unit area will *always* be produced. Second, if they also have lower unit transport costs per unit area of production, they will crowd out the competing products at any radius where rents are non-negative. However, if the higher (revenue minus production costs) crop also has higher transport costs, there exists a radius r' where rents are greater for the second commodity, and it alone is produced beyond this point. This is quite a plausible result, implying a *gravity* influence on the location of rural production. The model not only handles the usual supply/demand competition, but also the profit maximizing behaviour of the single landlord, who rents land to the crop owners with the greater ability to pay. For a cogent re-assessment of von Thünen's work, the reader is referred to Samuelson (1983). The exercise below should help the reader further to grasp the approach.

Exercise 2.1 Consider a city obtaining two agricultural commodities from its hinterland. With commodity *1* defined as that having the greater $(Q_j\, t_j)$ value, the data is given as:
$Q_1 = 200$ *tonnes/km*^{2}; $z_1 = \$295/tonne$; $t_1 = \$3.75/tonne$; $D_1 = 292250 - 100\, v_1$
$Q_2 = 250$ *tonnes/km*^{2}; $z_2 = \$256/tonne$; $t_2 = \$2.00/tonne$; $D_2 = 539000 - 150\, v_2$

(i) Assuming trial starting prices of $v_1 = 400$ and $v_2 = 320$, compute p_{01}, p_{02}, r' and r_m, as well as the total quantities S_j supplied.

(ii) Setting these supplied quantities in place of the quantities D_j in the demand functions, find the modified prices yielding Demand = Supply.

(iii) Using the modified prices from (ii), recompute steps (i) and (ii) once more, yielding a new set of prices. Are we converging?

(iv) In line with the classical 'cobweb' procedure of economics, substitute the relations (2.2) and (2.3) into the right-hand sides of (2.4) to obtain the supply functions. Invert these functions to obtain the inverse supply functions with the v_j's as functions of the S_j's. *Hint.* Substitute r'^2 from the first of (2.4) into the second of (2.4) and solve for v_2 as a function of S_1 and S_2. Using this function for v_2 in (2.3) for r_{01}, determine an expression for r', substitute it into the first of (2.4) and solve for v_1.

(v) Again using the starting prices $v_1 = 400$ and $v_2 = 320$, substitute these into the exogenous demand functions to determine the demands D_j. Using these quantities in place of the supplied quantities in the inverse supply functions, obtain the modified prices, and so on, until convergence (which is very rapid).

Urban Land Use Models. Three key contributors to the development of models to depict the urban land market are Alonso (1964) and Mills and Hamilton (1972). Rather than reproducing the original derivations from these texts, we build on Parable 3 in Mohring (1993) for a monocentric city, described more generally in Roy *et al.* (1998).

The classical assumptions are as follows: (i) all employment is concentrated at a single central point, the Central Business District (CBD), (ii) the city is situated on a featureless plain containing agricultural land of uniform fertility, (iii) each household contains one worker who commutes to the CBD each week day, (iv) there is just one mode of transport, (v) retail/service travel is taken as a constant amount for each household (and thus neglected), (vi) household re-location costs are negligible, (vii) all households are perfectly homogeneous, (viii) all income not devoted to land rent and travel is spent on one 'numeraire' good and (ix) each household spends the same amount on the dwelling (that is, the rents can be interpreted as *land* rents). The notation is as follows:

l = Distance (radial) of any point from CBD
t = Cost of travel per unit distance
I = Income per household
$L(l)$ = Size of housing lot occupied by a household at distance l
$X(l)$ = Numeraire good consumed by a household at distance l
$r(l)$ = Unit land rent at distance l
r^0 = Unit rent of agricultural land
Γ^L = Household utility weighting on land consumption at all points
Γ^X = Corresponding weighting on numeraire good

$$(\Gamma^L + \Gamma^X = 1)$$

$U(l)$ = Direct utility of household at distance l
$U^i(l)$ = Indirect maximum utility of household at distance l
U^r = Uniform utility realised at long run equilibrium by each household
$\lambda(l)$ = Lagrange multiplier on income constraint at distance l
N = Total (exogenous) number of households to be accommodated
R = Extreme (endogenous) outer radius of city

Assume that we have the following Cobb-Douglas utility function defined for any household at distance l in terms of land and the composite good

$$U(l) = \Gamma^L \log L(l) + \Gamma^X \log X(l) \tag{2.5}$$

which under the income constraint

$$X(l) + r(l)\, L(l) + t\, l = I \tag{2.6}$$

is maximized in terms of the quantities $X(l)$ and $L(l)$ given rents $r(l)$, yielding

$$X(l) = \Gamma^X [I - tl] \qquad L(l) = \Gamma^L [I - tl] / r(l) \tag{2.7}$$

after eliminating the Lagrange multiplier $\lambda(l)$ on (2.6). If the quantities in (2.5) are eliminated by substituting the utility maximizing values from (2.7), we obtain the *indirect utility function* in the form

$$U^i(l) = \Gamma^L \log \Gamma^L + \Gamma^x \log \Gamma^X + \log (I - t\, l) - \Gamma^L \log r(l) \tag{2.8}$$

Now, unless $U^i(l)$ is a uniform value at all values of l, households will have an incentive to move to a higher utility location and the system will not be in long run locational equilibrium. Thus, we impose a configuration of rents $r^*(l)$ to ensure that the same (unknown) utility level U^u is attained everywhere. By transposition of (2.8), we can obtain the equilibrium rent pattern $r^*(l)$ as

$$r^*(l) = \Gamma^L (\Gamma^X)^{\Gamma^X/\Gamma^L} (I - tl)^{1/\Gamma^L} \ \exp - (U^u / \Gamma^L) \tag{2.9}$$

As the rent reduces to the agricultural rent r^0 at the (unknown) periphery R of the city, this fact can be used to determine the uniform realised utility U^u by substitution into (2.9), yielding

$$U^u = \Gamma^L \log \Gamma^L + \Gamma^X \log \Gamma^X + \log (I - tR) - \Gamma^L \log r^0 \tag{2.10}$$

When (2.10) is substituted into (2.9), $\overline{U}$ is removed, yielding

$$r^*(l) = r^0 [(I-tl) / (I-tR)]^{1/\Gamma^L} \tag{2.11}$$

Upon substitution of (2.11) into the expression $L(l)$ for lot size in (2.7) to obtain the equilibrium lot size $L^*(l)$, noting that $[1 / L^*(l)]$ is the residential density at distance l, we may integrate to determine the exogenous peripheral radius R of the city in terms of the exogenous population N to be accommodated via

$$N = 2\pi \int_0^R [l / L^*(l)]\, dl \tag{2.12}$$

Although this integral can be derived analytically 'by parts', the resulting non-linear equation (2.12) must be solved iteratively for the peripheral radius R. This in turn allows the rent distribution to be obtained from (2.11) and the lot size and numeraire good distributions from (2.7). Also, it is more realistic to divide the population into separate income groups, competing for lots. If the maximum bid is always taken as successful, the lower income groups will have to be content with either small lots, inaccessible locations or both (Roy *et al.*, 1998).

In contrast to the analysis of Alonso and Mills, the above demonstration does not use geometric constructions, such as indifference curves, to illustrate the ideas. Instead, normal demand analysis via utility maximization is applied, as well as criteria for locational equilibrium. Nevertheless, the reader should refer to these fundamental references to enrich his understanding. At the same time, it should be noted that we assume that the landowners will supply land at the equilibrium rents $r^*(l)$. Thus, these rents are not determined via the usual short run price adjustment process between demand and supply, but rather as a consequence of the long run *locational* adjustments of households.

2.1.2 Agglomeration in Linear Markets for Single and Comparison Goods

One of the most interesting topics in spatial economics is identification of the tendency for activities to concentrate at discrete nodes. This phenomenon can be examined as follows, both for a single stop to purchase a single homogeneous good and for multiple stops to choose between heterogeneous goods. At the same time, we can contrast competitive equilibria with welfare-maximizing solutions.

Single Stop to Purchase a Homogeneous Good. This is the case considered in Hotelling (1929). We list the following assumptions on the nature of the linear market: (i) the linear market is of unit length, (ii) customers are spread uniformly along the line, (iii) two identical firms are present, located somewhere along the line, (iv) each customer makes one visit to the *closer* firm along the line and (v) each firm can costlessly change its location. The questions asked by Hotelling are 'does a long run locational equilibrium exist and is it stable'? He answered 'yes' to both questions. From a 'socially' optimal point of view, it is easy to see that the least average travel distance of *0.125* is consumed if one firm locates at *0.25* and the other at *0.75* - each firm would also have identical sales. But, what about stability? For instance, if the left-most firm moved from *0.25* to *0.74999*, just to the left of the right-most firm, it would claim *3/4's* of the market in the short run. On the other hand, the right-most firm could leapfrog over the left-most firm and itself claim almost *3/4's* of the market. This leapfrogging would continue until both firms are concentrated at the *0.5* point where (a) they would both have equal market share and (b) neither firm could unilaterally change it's position without becoming worse off. These are exactly the criteria for a stable *Nash* locational equilibrium. Thus, we have a natural reason for *concentration* in spatial markets without having to invoke non-linearities such as scale economies. Note that, this stable long run competitive equilibrium is not socially optimal - the average access distance for customers would double to *0.25* compared to the situation of firms locating at the quarter points. Although this example is idealised, it does identify a possible role for *locational policy* - for instance, rent incentives could be offered to a single firm at each quarter point to induce it to stay there. Note that, the 'gravity' effect is restricted to the customer choosing the closer firm.

Multi-Stop Trips for Comparison Goods. The ideas of Hotelling were generalised by Eaton and Lipsey (1979) for linear cities with multiple facilities accessed by multi-stop trips. Sharpe and Roy (1985) extended Eaton and Lipsey's analysis in a topological sense to *rings* of facilities accessed circumferentially. In this exposition, we concentrate on Eaton and Lipsey's scheme. Whilst each facility is taken as equally attractive on the average, individual customers are assumed to *perceive* the necessity of visiting more than one store on a particular trip, minimizing their *combined* access distance. The results for the case of *2-stop trips* may be summarised as follows, assuming that, after completing the comparison trip, each customer makes a delivery order to *one* of the facilities visited with equal *(50%)* probability: (a) peripheral firms must be grouped in a least a *triplet,* (b) no single grouping contains more than *four* firms, (c) no firm can be unpaired, (d) the market area of each grouping extends half the distance to the neighbouring group on either side, (e) no firm's whole market segment can be less than any other firm's half-market segment, (f) the location of triplets and quartets is uniquely determined at the midpoints of the markets which they serve and (g) the location of any pair that is neither the second nor the penultimate group is not unique within its interval. This is a strong result, with the clustering being much less surprising than for the single stop case. As an example, consider a linear city needing to accommodate *6* facilities. Again, we are interested in the existence, uniqueness and stability of competitive equilibria, as well as in the social optimum. For the social optimum, we have three clusters of stores, with two stores per cluster, located at the *{1/6, 1/2, 5/6}* points, as shown in Fig. 2.2.

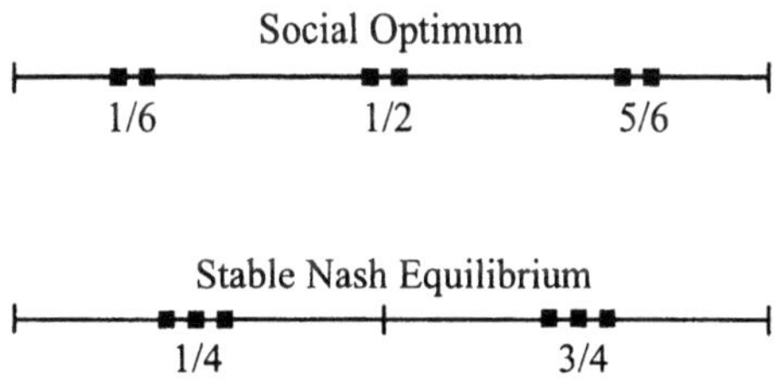

Fig. 2.2 Store Locations in Linear Market (2 stops)

The average distance travelled is just *1/12*, with each customer satisfying their requirements within a single cluster pair. As the market share (ie. profit) of each firm is identical here, it also represents a market equilibrium. However, as can be deduced, it is not stable. For instance, if one of the firms, say at the *1/6* point, decided to move just to the left of the *{1/2}* point, it would, as before, capture all its previous customers from *0* to *1/3*, plus the customers between *1/3* and *1/2*, increasing its market length from *1/3* to *1/2*. This is clearly an unstable situation, inducing a further shift of a firm at the *1/2* point to try to regain its position. In fact, Eaton and Lipsey showed that the only *stable* competitive equilibrium occurs for *triplets* at the *1/4* and *3/4* points, where customers now travel a distance of *1/8* on the average. The reader can experiment with moves from different players to confirm that any such unilateral move would make the mover worse off.

Although the assumptions leading to the Hotelling and Eaton and Lipsey models are rather idealised, they are nevertheless not implausible in principle. They appear to provide a 'natural' spatial explanation of concentration of economic activities. Of course, by neglecting things such as congestion, they are unable to identify any countervailing 'centripetal' tendencies.

Exercise 2.2 Consider a linear market of unit length along which demand is uniformly distributed. Assuming that customers visit *2* facilities on a comparison trip and four facilities are to be located, proceed as follows:

(i) Verify that both the social optimum and a competitive optimum occur when pairs of facilities are located at the *(1/4)* and *(3/4)* points. Evaluate the average distance travelled per customer.

(ii) Starting with configuration (i), move one of the facilities at the *(1/4)* point a small distance δ to the left of the *(3/4)* point. Compute the market share of the moved facility and show that it has (temporarily) improved its position compared with its *(1/4)* share in (i). What does this imply for the *stability* of the competitive equilibrium?

(iii) Consider all facilities to be concentrated at the centre *(1/2)* of the line, thus all having equal market share and constituting a competitive equilibrium. Move one facility a distance δ to the left of the centre. Does its market share improve? What does this imply for the stability of this equilibrium? Also, compute the average distance travelled.

2.1.3 Linear Model of Production Location and Transport Flows

In this model, the regional location of production is adapted to the locations of demand for a single commodity such that the total transport costs in the system are minimized. This classical 'transportation problem of linear programming' was developed by Koopmans (1949). We note that the demand is given exogenously, and may arise from consumer demand plus any demand for the commodity as intermediate inputs to the production of other commodities. The model is short run, in that production capacity for the commodity is given exogenously in each region. It is assumed that the price of the commodity is uniform over all regions. Also, we have *uniform delivered* pricing, where the producers pay the transport delivery costs of the commodity. This implies that the conventional profit maximizing objective of the competitive firm is replaced by minimization of production plus transport costs. This is taken as the total production costs over the entire system, which is extrapolated from the individual cost minimizing behaviour of each individual firm. This involves not only the assumption of identical firms within each region, but of some sort of 'coordination' between regions to share the available demand. The model is linear in two senses (a) production costs are linear with respect to quantity produced and (b) transport costs are linear with respect to *quantity* shipped (non-linearities are acceptable with respect to shipping distances). The following variables are introduced:

K_i = Production capacity of good in region i
D_j = Demand for good in region j
t_{ij} = Transport cost per unit of good shipped between i and j
c_i = Production cost per unit of good in region i
q_{ij} = Unknown quantity of goods shipped between i and j
$q_i = \Sigma_j q_{ij}$ = Unknown total production in region i ($\leq K_i$)

The objective is to determine the shipment pattern q_{ij} which minimizes the total cost C of production and goods transport, expressed as

$$C = \Sigma_{ij} [c_i + t_{ij}] q_{ij} \tag{2.13}$$

Assuming that all demand is taken up, the following constraints are applied at each demand region j

$$\Sigma_i q_{ij} = D_j \tag{2.14}$$

At each producing region i, the actual production $q_i = (\Sigma_j q_{ij})$ must not exceed the short run productive capacity K_i, yielding the constraints

$$\Sigma_j q_{ij} \leq K_i \tag{2.15}$$

The system (2.13) to (2.15) represents a standard *linear program*, with a linear objective function (2.13) and linear constraints (2.14) and (2.15). Associated with the constraints (2.14) are Lagrange multipliers η_j, and with the inequality constraints (2.15) Lagrange multipliers λ_i (zero when inactive). The result is a unique pattern of flows q_{ij}.

In this class of problem, the Lagrange multipliers have useful interpretations. By definition, the Lagrange multipliers η_j on the demand constraints represent the reduction in total transport costs which would result upon a unit increase of demand in region j - thus, if there are any regional demand policies being assessed by the analysis, those regions where η_j is greatest would be preferred candidates for any induced increase in demand. Similarly, regions with the highest non-zero value of the capacity multiplier λ_i would result in the greatest cost reduction for a unit increase in capacity.

Finally, the above problem has become known as *the transportation problem in linear programming*. As shown in the path-breaking paper (Evans, 1973), if the *objective* (2.13) is introduced as a *constraint* on an entropy maximizing problem, together with (2.14) and (2.15), in terms of distinguishable flow units, the solution of the entropy problem would approach asymptotically that of the linear program above as the Lagrange multiplier on the new cost constraint approaches infinity. This fundamental result, which holds in a wide variety of other supply and demand contexts, allows us to interpret the deterministic problems of economics (for example, cost minimization or profit maximization for the competitive firm) as special limiting cases of analogous probabilistic entropy maximization problems, where the deterministic objective of the former is replaced by a constraint on the latter evaluated from observations at a given base period. It is the generalisation of this result, enhanced by interpretations of the ideas of Smith (1990), that enable us to conceptually link spatial interaction models and microeconomic models, the main purpose of this book. The 'theory' is encapsulated in the form of the cost or profit (or utility) constraint, which the entropy model allows to be satisfied by base period observations, yielding an empirically-based enhancement of the original deterministic 'theory'.

2.1.4 Imperfect Competition in Linear Space

In the competitive model, the number of actors is assumed to be so large that no single firm or individual can influence market outcomes. This means that both firms and individuals are 'price-takers'. Although such markets may exist for relatively homogeneous commodities with a large number of participating firms, it is an idealised abstraction for many modern skill-based enterprises. In these cases of *imperfect* competition, each firm is assumed to (i) have a level of understanding of the response of demand to price changes and (ii) have some level of perception of rival firms' behaviour. In common with Beckmann and Thisse (1986), we believe that key characteristics of spatial markets can best be appreciated via elegant analyses which reduce a problem to its essentials. In the following, we rely on Beckmann and Thisse for two main cases, monopoly *(1 firm)* and duopoly *(2 firms)*, the simplest case of oligopoly (several firms).

Spatial Monopoly in a Linear Market. Consider, as shown in Fig. 2.3, a single firm located at distance x* from the left end of a linear market of unit length. Let the demand entities be uniformly distributed along the line and let *mill pricing* prevail, where the demand firms pay the transport costs. Let c be the unit cost of production, t the unit transport costs per unit distance, p the (endogenous)

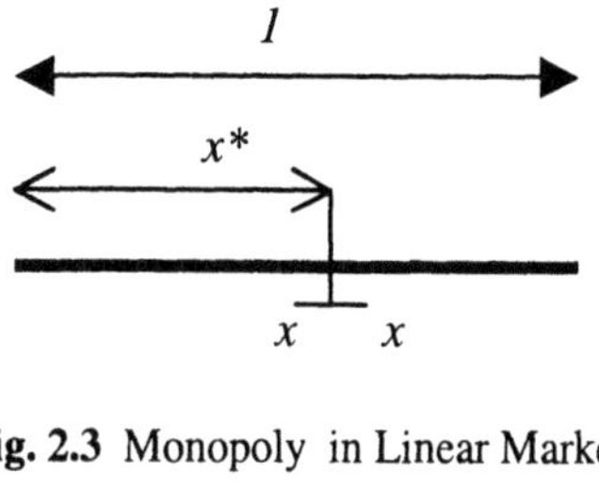

Fig. 2.3 Monopoly in Linear Market

unit price of the commodity and x the distance from the producing firm to any point along the line, right or left. The total profit Π is maximized in terms of the price p, assuming a linear demand function $(1 - p - tx)$, with $(p - c)$ being the profit per unit of demand. In this *short run* problem, the location of the firm is taken to remain unchanged at point x^*. Thus, the objective is expressed as

$$\Pi = \int_0^{x^*} (p - c)(1 - p - tx)\, dx + \int_0^{(1 - x^*)} (p - c)(1 - p - tx)\, dx \tag{2.16}$$

The integrals are evaluated to yield

$$\Pi = (p - c)\{ (1 - p) - t(x^{*2} - x^* + 1/2) \} \tag{2.17}$$

If (2.17) is maximized in terms of p, we obtain the unique equilibrium price as

$$p = [1 + c - t(x^{*2} - x^* + 1/2)] / 2 \tag{2.18}$$

which is seen to be a maximum via the sign of the second derivative.

The price p is the monopolist's price at the specified location x^*. However, in the *long run*, the firm may consider changing its location to further increase its profits. In order to determine the optimal location, we can differentiate the profit expression (2.17) with respect to the point of location x^*, yielding the following condition for a maximum

$$-(p - c)\, t[2x^* - 1] = 0 \tag{2.19}$$

which gives the optimal value of x^* equal to 0.5, as expected. If this value is substituted into (2.18), the short run equilibrium price at the long run equilibrium location comes out as

$$p = [1 + c - t/4] / 2 \tag{2.20}$$

We note that, although the producer does not pay the transport costs, its price is negatively affected by transport costs because of the dampening effect on demand.

Spatial Duopoly in a Linear Market. Consider that we have *2* Cournot firms (Intriligator, 1971), who compete for demand along a linear market, but do not have the knowledge to anticipate their rival's decisions. The market is depicted in

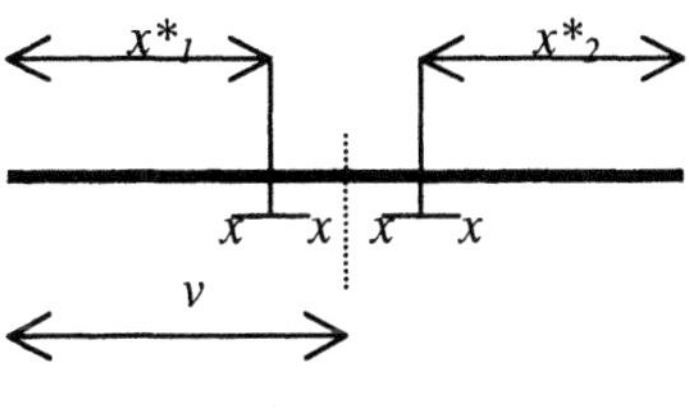

Fig. 2.4 Duopoly in Linear Market

Fig. 2.4, in which firm *1* is located at point x^*_1 and firm *2* at point x^*_2. It is assumed that the unit production costs of the firms are c_1 and c_2, which differ by less than the maximum transport cost - otherwise, one firm would be dominant. In the short run, given the locations x^*_1 and x^*_2, we wish to obtain the equilibrium prices p_1 and p_2, as well as the market share v and $(1 - v)$ of firms *1* and *2* respectively. Now, as for the monopoly case above, the profit expression for firm *1* can be written as

$$\Pi_1 = \int_0^{x^*_1} (p_1 - c_1)\,(1 - p_1 - tx)\;dx \;+\; \int_0^{(v-x^*_1)} (p_1 - c_1)\,(1 - p_1 - tx)\;dx \qquad (2.21)$$

with a similar expression for firm *2*. Integration yields the results

$$\begin{aligned}\Pi_1 &= (p_1 - c_1)\;\{(1 - p_1)\,v - t(x^{*2}_1 - vx^*_1 + v^2/2)\,\}\\ \Pi_2 &= (p_2 - c_2)\;\{(1 - p_2)\,(1-v) - t[x^{*2}_2 - (1-v)x^*_2 + (1-v)^2/2]\,\}\end{aligned} \qquad (2.22)$$

Assuming that demand firms (or consumers) select the firm which has the lower *delivered* price, the break-even point v is evaluated via the condition

$$p_1 + t(v - x^*_1) = p_2 + t(1 - v - x^*_2) \qquad (2.23)$$

yielding v as

$$v = [(p_2 - p_1) + t(1 + x^*_1 - x^*_2)]\,/\,2t \qquad (2.24)$$

allowing it to be eliminated from the profit expressions (2.22). The next step is to partially differentiate the first of (2.22) with respect to p_1 and the second with respect to p_2, realising that v is a linear function of both p_1 and p_2. The simultaneous presence of *both* prices p_1 and p_2 in the expression for v, and thus in the profit expressions for *both* Π_1 and Π_2 in (2.22), defines the *strategic interdependence* between firms which is typical for oligopoly situations.

The mathematics is more complicated by an order of magnitude than for the monopoly problem, and will not be shown here in detail. Suffice it to say that the substitution of v, differentiation of the profit functions and setting them to zero yields two simultaneous quadratic equations for p_1 and p_2, which can be solved numerically by successive substitution. The value of v is then directly determined from (2.24). Note that, the above analysis may be adapted to *uniform delivered pricing,* where the producers pay the transport costs.

With the short run problem solved, the next step is, with the function for v substituted from (2.24), to solve the *long run* locational equilibrium problem by maximizing the profits Π_1 and Π_2 in (2.22) with respect to the firm locations x^*_1 and x^*_2 respectively. It remains as an exercise for the reader to verify, that when the two firms are identical, that is the production costs c_1 and c_2 are equal to a constant value c, the optimal long run locations for the two firms are at the *(1/4)* and *(3/4)* points respectively. These are the locations where total transport costs are minimized, also corresponding to a social optimum when transport cost minimization is taken as the social welfare objective.

The reader will appreciate from the above that spatial oligopoly problems are difficult to solve, even for the simplest case of two 'myopic' Cournot firms in simple linear markets. Similar challenges arise for the case of monopolistic competition (Dixit and Stiglitz, 1977). At the same time, with oligopolistic or monopolistic competition behaviour being increasingly relevant in many niche markets for high value-added commodities and services, the solution of these problems is an important challenge. The use of entropy in later chapters allows a clear stochastic representation of the deterministic solutions demonstrated above.

2.2 From Theory to Applications

2.2.1 National, Interregional and Multi-Regional Input-Output Models

Multi-regional and interregional input-output modelling is an extension of non-spatial (national) input-output modelling developed by Leontief (1951), Leontief *et al.* (1953) and Chenery and Clark (1959). Other key contributors are Polenske (1980) and Hewings (1985). To aid understanding of the regional extensions of the technique, the basics of national non-spatial input-output are first described.

National I-O Model. Consider a national economy. Then, let x_{ij} be the quantity of intermediate input of economic sector i going into the process of production of economic sector j. Let x_k be the total production (output) of any economic sector k. The *usage* of the total output x_i of a sector i is either for intermediate inputs to other sectors j or for direct consumption in quantities y_i, defined as *final demand*. This final demand may either be consumed within the country or exported. Then, the following *transaction balance* relations apply universally for each sector i

$$x_i = \sum_j x_{ij} + y_i \tag{2.25}$$

In order to derive the demand x_{ij} for inputs i in each sector j of the economy, we turn to the classical theory of the firm, as given for example in Intriligator (1971). Given a set of inputs, it is postulated that there is a unique maximum output. This technological relationship between a set of inputs and the maximum output is called the *production function.* For the case of the *input-output production function*, let a_{ij} be the minimum amount of input i needed in the overall production process for a unit of output in sector j. The production function is given as

$$x_j = min_i \{x_{ij} / a_{ij}\} \tag{2.26}$$

in which min_i denotes the minimum over all input sectors i used for the output of sector j. In other words, the output is constrained *independently and linearly* by the amounts available of each and every input and its relative transformation level a_{ij} into output. This implies that there is *zero substitution* in an input-output economy. A further property of (2.26) is that it is *non-differentiable*, complicating the mathematics for determining the input demand functions in terms of profit maximizing firms (sectors). If that total profit is defined as Π and the unit prices of the goods are defined as p_i, then we can write the profits as total sales revenue minus input costs in the form

$$\Pi = \Sigma_j p_j x_j - \Sigma_{ij} p_i x_{ij} \tag{2.27}$$

Now, if (2.26) is substituted for the output x_j in (2.27), the profit expression is entirely in terms of the input transactions x_{ij}. If the resulting expression is differentiated with respect to x_{ij} and equated to zero, we can, in principle, solve for the profit-maximizing inputs x_{ij} as the *input demand functions*. However, as shown by Chenery and Clark (1959), this procedure is complicated by the non-differentiability of (2.26). Nevertheless, it is possible to show that the input demand functions come out simply as the proportional relations

$$x_{ij} = a_{ij} x_j \tag{2.28}$$

However, the elegance of this result is illusory. Firstly, prices have disappeared from (2.28). Secondly, despite eliminating the output quantities x_j from (2.27) by substitution via (2.26), they have re-appeared in the input demand relations (2.28)! In other words, we do not know the profit-maximizing input quantities absolutely, but just as a ratio of the *unknown* output quantities x_j. This is in contrast to the classical case, where input demand quantities are determined absolutely, with output quantities being obtained endogenously by substitution of the input demand quantities into the production function, yielding the *supply function*. However, if we take the final demand y_i in (2.25) as being provided exogenously from final demand projections, then the substitution of (2.28) into (2.25) eliminates the input demand quantities to give expressions entirely in terms of the outputs x_i and the known final demands y_i, yielding

$$x_i = \Sigma_j a_{ij} x_j + y_i \tag{2.29}$$

which is transposed, put in matrix form and solved for the output vector $\boldsymbol{X}$ via

$$\boldsymbol{X} = (\boldsymbol{I} - \boldsymbol{A})^{-1} \boldsymbol{Y} \tag{2.30}$$

In (2.30), $\boldsymbol{I}$ is the identity matrix and ' $^{-1}$ ' the symbol for matrix inversion. Thus, for any projected values of final demand $\boldsymbol{Y}$ and current values in the input-output matrix $\boldsymbol{A}$, the solution (2.30) yields the resulting output $\boldsymbol{X}$. The corresponding inter-sectoral transactions x_{ij} may then be obtained by substitution into (2.28).

Though the above analysis is relevant for a single country, there is no indication how the flows are subdivided between the regions of the country. Such estimates are vital in the implementation of regional policy, including transportation investment policy. In fact, we are interested to determine *interregional* transactions x_{ij}^{rs}, from sector i in region r to sector j in region s, or, at least *multi-regional* transactions x_i^{rs} of sector i between regions r and s.

Interregional I-O Model. Examining first *interregional* analysis, the transaction equations (2.25) are generalised to

$$x_i^r = \sum_{js} x_{ij}^{rs} + y_i^r \tag{2.31}$$

where x_i^r and y_i^r are the output and final demand respectively of sector i in region r. Similarly, the input demand equations (2.28) now become

$$x_{ij}^{rs} = a_{ij}^{rs} x_j^s \tag{2.32}$$

where the *interregional input-output coefficients* a_{ij}^{rs} denote the amount x_{ij}^{rs} of input of sector i in region r being input into sector j in region s per unit of output x_j^s of sector j in region s. If the relations (2.32) are substituted for x_{ij}^{rs} into (2.32), a result analogous to (2.30) emerges, but of much higher dimensionality. With computers becoming more and more powerful, the solution of such large systems is no great problem. However, the acquisition of sufficient survey data to allow evaluation of the full set of interregional coefficients a_{ij}^{rs} is a considerable hurdle to implementation of a full interregional analysis.

Multi-Regional I-O Model. In multi-regional modelling, we merely need I-O coefficients a_{ij}^s, which do not distinguish the regions of origin of the inputs. Thus, they are a true representation of the *technology* operating in region s for sector j. The proportionality relations (2.32) here become

$$x_{ij}^s = a_{ij}^s x_j^s \tag{2.33}$$

where the x_{ij}^s represent the total inputs to sector j of goods from sector i in region s. As described in Polenske (1980), commodities from each origin region are put into a supply *pool* in the destination region, from which they are distributed as inputs to final demand and all other sectors in constant proportions relatively from each origin region. These proportions are expressed as trade coefficients t_i^{rs} via

$$t_i^{rs} = x_i^{rs} / \{\sum_r x_i^{rs}\} \tag{2.34}$$

where x_i^{rs} represent the total flows of sector i goods between regions r and s. Such flows are usually available from commodity shipment data. From the above, it is straightforward to show that the multi-regional problem is given in matrix form as

$$\boldsymbol{X} = (\boldsymbol{I} - \boldsymbol{T}\,\boldsymbol{A})^{-1}\, \boldsymbol{T}\boldsymbol{Y} \tag{2.35}$$

Once (2.35) is solved for the output vector $\boldsymbol{X}$, the inter-sectoral transactions x_{ij}^s within each region can be obtained. When these transactions are summed over all the sectors j to be supplied and final demand y_i^s is added, we obtain the total amount of sector i goods available for use in region s, corresponding to the denominator of (2.34). Use of the trade coefficients in (2.34) then allow the flows x_i^{rs} of the goods of each sector i between each pair of regions to be evaluated.

The RAS Approach for Updating Tables. The RAS procedure was developed by Stone and others in Cambridge University (1963) for updating input-output tables via updating of the associated flow matrices. For example, one may want to obtain a table $\boldsymbol{A}^{ss}$ for a particular region s by updating the national input-output table $\boldsymbol{A}^{nat}$. At the regional level, it is assumed (omitting the regional identifier s) that one has the gross output vector $\boldsymbol{x}$, as well as the row sums $\boldsymbol{u}$ of all intermediate input transactions (that is, output minus final demand for each sector) and the column sums $\boldsymbol{v}$ of these same transactions (that is, output minus value added minus imports). Thus, a priori, we make the first estimate $\boldsymbol{X}^0$ of the regional transactions matrix as

$$\boldsymbol{X}^0 = \boldsymbol{A}^{nat}\,\boldsymbol{x} \qquad (2.36)$$

The Cambridge approach is motivated here in terms of information theory, as demonstrated by Batten (1983) and described in general in Chapter 1. In order to determine the revised transactions matrix $\boldsymbol{X}$ which has the correct row and column sum vectors $\boldsymbol{u}$ and $\boldsymbol{v}$ respectively, as well as being as *close* in an information sense to the prior transactions matrix $\boldsymbol{X}^0$, we set up the following minimum information objective $\boldsymbol{Z}$, to be minimized in terms of x_{ij}

$$\boldsymbol{Z} = \Sigma_{ij}\, x_{ij} \log\,(x_{ij} / x_{ij}^{\;0}) \qquad (2.37)$$

under the following row and column sum constraints

$$\Sigma_j\, x_{ij} = u_i \;;\quad \Sigma_i\, x_{ij} = v_j \qquad (2.38)$$

This problem is solved for the fully consistent transactions matrix $\boldsymbol{X}$ using the well-known iterative bi-proportional adjustment algorithm, whereupon the values a_{ij} of the updated regional input output matrix $\boldsymbol{A}^{ss}$ are then given simply as

$$a_{ij} = x_{ij} / x_j \qquad (2.39)$$

The RAS procedure may also be used for updating either transactions or input-output tables over time, assuming that estimates are available of future gross outputs and the row and column sums of the intermediate transactions. In fact, knowledge of gross outputs, final demand, value-added, external imports and external exports turn out to be quite sufficient to enable the right-hand sides of (2.38) to be evaluated.

2.2.2 Spatial Price Equilibrium

Spatial price equilibrium modelling was developed by Samuelson (1952a) and Takayama and Judge (1964, 1971) to examine static equilibrium in a competitive commodity market where the interregional transport costs are significant with respect to the market price of the commodity. This definition includes most agricultural commodities and many raw materials.

In this brief discussion of the approach, it is assumed that we have demand and supply functions expressing quantities in terms of prices. Let there be a set of supply nodes i at which we have a vector $\boldsymbol{s}$ of supply functions $s_i(\boldsymbol{\pi})$ which are potentially functions of all the supply prices π_i gathered within a supply price vector $\boldsymbol{\pi}$, yielding the vector relation

$$\boldsymbol{s} = \boldsymbol{s}(\boldsymbol{\pi}) \tag{2.40}$$

Let there also be a set of demand nodes j at which we have a vector $\boldsymbol{d}$ of demand functions $d_j(\boldsymbol{\rho})$ which can be functions of all the demand prices ρ_j within a demand price vector $\boldsymbol{\rho}$, expressed via the following vector relation

$$\boldsymbol{d} = \boldsymbol{d}(\boldsymbol{\rho}) \tag{2.41}$$

Let the unit transport costs between each market pair be grouped in a matrix $\boldsymbol{c}$. Then, if these unit costs are flow-dependent and the unknown flows are grouped into a matrix $\boldsymbol{Q}$, $\boldsymbol{c}$ is expressed as the following function

$$\boldsymbol{c} = \boldsymbol{c}(\boldsymbol{Q}) \tag{2.42}$$

Consider a pair (i,j) of supply and demand markets, between which *mill pricing* applies, that is, the demand firms pay the transport and handling costs c_{ij}. Then, if the unknown flows are grouped into a matrix $\boldsymbol{Q}$, equilibrium is characterised at prices $\boldsymbol{\pi}^*$ and $\boldsymbol{\rho}^*$ by the following relations

$$\pi_i^* + c_{ij} \begin{cases} = \rho_j^* & \text{if } Q_{ij}^* > 0 \\ \geq \rho_j^* & \text{if } Q_{ij}^* = 0 \end{cases} \tag{2.43}$$

This implies that goods will only be traded at equilibrium between any two markets i and j if the supply price plus the transport costs equals the demand price. If the supply price plus transport cost is greater than the demand price between any pair of markets i and j, then there will be no goods traded between this market pair. In addition, it is necessary to satisfy the flow consistency conditions

$$s_i^* = \Sigma_j Q_{ij}^* \quad ; \quad d_j^* = \Sigma_i Q_{ij}^* \tag{2.44}$$

which represent market clearing at equilibrium.

The objective is to solve the combined system (2.40) to (2.44). At the time when the initial formulation by Takayama and Judge (1964) was made, the problem was re-cast into a non-linear mathematical programming framework, allowing quite large problems to be handled. More recently, variational inequality solutions, such as in Nagurney (1993), have been efficiently implemented for this class of problem, and imperfect competition has been introduced. Nevertheless, the problem is still given as deterministic. In Chapter 3, probabilistic approaches to this class of problem are discussed, allowing non-price factors and gravity influences to influence the trading patterns.

2.3 Some New Directions

2.3.1 Increasing Returns and Monopolistic Competition

Although the relevance of increasing returns in manufacturing at the level of the individual firm started to be considered in the context of imperfect competition by economists such as Romer (1986) and Arthur (1994) and his colleagues, the use of such concepts in international trade by Helpman and Krugman (1985) first introduced the spatial dimension. In this 'new economic geography', applications at the regional level were made by Fujita *et al.* (1999). There will be no attempt in this book to place such recent work in a probabilistic spatial interaction framework. Nevertheless, the general concepts introduced here in Chap. 3 could, in principle, be applied to this task. Instead, we sketch the key model framework introduced in Chap. 4 of Fujita *et al.* (1999) to give the reader an appreciation of the ideas involved. It represents a spatial version of the Dixit and Stiglitz (1977) model of monopolistic competition, as the first step towards a more realistic version.

Consider a two-sector economy, having a homogeneous competitive agricultural sector and a manufacturing sector composed of differentiated goods idealised into a continuous product space.

Consumer Model. Identical consumers consume the two goods in terms of the following Cobb-Douglas utility function

$$U = M^{\mu} A^{1-\mu} \tag{2.45}$$

where M is an index for consumption of the diverse manufactured good, A is the consumption of the agricultural good and μ a constant, denoting the expenditure share of manufactured goods. Define M as the following CES (Constant Elasticity of Substitution) function across the continuous 'range' n of varieties of goods

$$M = \left[\int_0^n m(i)^{\rho} \, di \right]^{1/\rho} \qquad 0 < \rho < 1 \tag{2.46}$$

where ρ denotes the reluctance for variety in manufactured goods, with more variety being preferred as $\rho \to 0$. With a given income Y, a price p^A for the agricultural good and $p(i)$ for the manufactured goods, the utility in (2.45) is maximized according to the budget constraint

$$p^A A + \int_0^n p(i)\, m(i)\, di = Y \tag{2.47}$$

In the two-step procedure for solving this problem, the first step involves minimization of the cost required to produce a given value M of the composite manufactured good, represented via solution of

$$\min \int_0^n p(i)\, m(i)\, di \;\text{ subject to }\; [\int_0^n m(i)^\rho\, di\,]^{1/\rho} = M \tag{2.48}$$

which yields the result

$$m(j) = M\, p(j)^{1/(\rho-1)} / [\int_0^n p(i)^{\rho/(\rho-1)}\,]^{1/\rho} \tag{2.49}$$

The price index G for manufactured products, as the expenditure per unit quantity is expressed as

$$G = [\int_0^n p(j)\, m(j)\, dj\,]/M = [\int_0^n p(i)^{1-\sigma}\, di\,]^{\,1/(1-\sigma)} \tag{2.50}$$

where σ is the elasticity of substitution $1 / (1 - \rho)$ between any two varieties. Now, with the help of (2.50), the demand relation $m(j)$ in (2.49) is given as

$$m(j) = M\, [p(j) / G]^{-\sigma} \tag{2.51}$$

The second step is to divide income into that spent on agriculture and that spent on manufactured goods, maximizing the utility (2.45) under the budget constraint (2.47) expressed as

$$G M + p^A A = Y \tag{2.52}$$

yielding the following consumer demand functions

$$A = (1 - \mu)\, Y / p^A \quad ; \quad m(j) = \mu\, Y\, p(j)^{-\sigma}\, G^{(\sigma-1)} \tag{2.53}$$

Finally, the indirect utility function gives (2.45) in terms of income and prices as

$$U = \mu^\mu (1 - \mu)^{1-\mu}\, Y\, G^{-\mu}\, (p^A)^{-(1-\mu)} \tag{2.54}$$

Note that, the range of goods n is endogenous in this problem.

Analysis for Several Regions. In order to avoid an explicit modelling of product flows, the 'iceberg' transport assumption of Samuelson (1952b) is used. If the consumers are assumed to pay the transport costs (*fob* pricing), then with a selling price p_r^M at producing region r for a manufacturing variety, the consumer price (*cif* price) p_{rs}^M for a unit of the same good when sold at region s is given as

$$p_{rs}^M = T_{rs}^M p_r^M \tag{2.55}$$

where an expansion factor T_{rs}^M (greater than unity) is assumed to allow for the effect of transport costs. Also, only the proportion $(1 / T_{rs}^M)$ of the original shipment is taken to arrive at the destination, the rest 'melting' on the way. Thus the composite price index G_s at each consumption region s is generalised from (2.50) for the system of R regions in the form

$$G_s = [\sum_{r=1}^{R} n_r (p_r^M T_{rs}^M)^{1-\sigma}]^{1/(1-\sigma)} \tag{2.56}$$

This allows (2.51) to be generalised to yield the total sales q_r^M for a variety at r to

$$q_r^M = \mu \sum_{s=1}^{R} Y_r (p_r^M T_{rs}^M)^{-\sigma} G_s^{(\sigma-1)} T_{rs}^M \tag{2.57}$$

where the final T_{rs}^M factor corrects for the amount lost en route.

Behaviour of Producers. Assuming the agricultural goods are traded in a perfectly competitive market under constant returns, we concentrate here on the manufactured goods. Economies of scale are taken to exist at the variety level of the manufactured goods. Assume identical technology for each variety in each location, with labour as the only input, with fixed costs F and marginal costs c^M, with the required labour input l^M expressed as

$$l^M = F + c^M q^M \tag{2.58}$$

Then, with manufacturing wages denoted by w_r^M, the unit profit π_r is written as

$$\pi_r = p_r^M q_r^M - w_r^M (F + c^M q_r^M) \tag{2.59}$$

Under the assumption of *monopolistic competition,* the inverse demand function (2.57) is substituted for q_r^M in (2.59). If the resulting expression is differentiated with respect to the price p_r^M, the equilibrium values emerge as

$$p_r^M = c^M w_r^M \sigma/(\sigma-1) \tag{2.60}$$

The resulting *short run* equilibrium quantities q_r^M can be obtained by direct substitution into (2.57). When we consider the *long run*, the relative profit π_r is assumed to adjust to zero in each region r, yielding the equilibrium ouput q^*, constant in each region as

$$q^* = F(\sigma - 1)/c^M \tag{2.61}$$

from which the equilibrium labour supply is readily computed.

The above yields a very revealing demonstration of the use of monopolistic competition and product variety to encompass scale economies at the firm level in a spatial market. In principle, this modelling framework can be extended by inserting the above deterministic theory into a probabilistic framework, as illustrated in the next chapter. At the same time, the resulting interregional flows can be determined.[1]

2.3.2 Knowledge Exchange and Space

The increasing importance of the 'knowledge economy' in industry has been emphasised by authors such as Kobayashi *et al.* (1991). In this brief note, we concentrate on knowledge networks among individuals, who communicate by electronic means, as well as enjoying different frequencies of face-to-face contacts, drawing on an elegant analysis by Beckmann (1994).

Consider a group of academics $i, j, k = 1, \ldots n$ in a particular field who both work on their own or in collaboration with another. For researcher i, the time (effort) devoted to these are x_{ii} working alone and x_{ij} in collaboration with researcher j. The output is characterised by a production function

$$f_{ij}(x_{ij}, f_{ji}) = b_{ij} f_{ij}(x_{ij}, x_{ji}) \tag{2.62}$$

if we standardise such that $f(1, 1) = 1$, yielding b_{ij} as the output attributable to i from unit contributions by i and j related to their affinity and complementarity. We specify f to be symmetric, that is, $f(y, z) = f(z, y)$. Also, let $b_{ij} = b_{ji}$ be half the total output from the collaboration. In particular, the production function is assumed to take the Cobb-Douglas form

$$f = \sqrt{(x_{ij}\, x_{ji})} \tag{2.63}$$

Also, each individual has a time budget c_i constrained via

$$\Sigma t_{ij}\, x_{ij} \leq c_i \tag{2.64}$$

where t_{ij} is the time spent to achieve one unit of effective collaboration between i and j. When $t_{ii} = 1$ for working alone, collaboration at a distance implies

$$t_{ij} = 1 + a\, r_{ij} > 1 \tag{2.65}$$

---

[1] Since the preparation of this chapter, the path-breaking work of Fujita and Thisse (2002) has appeared. This work has profound implications for long run locational models.

where r_{ij} is a representation of the relative distance between researchers i and j and a is a measure of the required relative trip frequency. In all cases, we assume that e-mail type interaction must be complemented by a certain frequency of face-to-face contacts - otherwise, 'distance' would be more cultural than physical.

Considering the case of individual pair-wise contacts, in which individual i firstly chooses the collaborator j with the greatest benefit/cost ratio via

$$b_{ij} / t_{ij} = max_k \ (b_{ik} / t_{ik}) \tag{2.66}$$

The contact will occur if j finds the contact better than working alone, that is, if

$$(b_{ji} / t_{ji}) > b_{jj} \tag{2.67}$$

Now, the focus shifts to the allocation of effort x_{ij} to achieve a system optimum for all individuals. The objective becomes

$$\max\ x_{ij} \ \Sigma_{ij} \, b_{ij} \ \surd(x_{ij} \, x_{ji}) \tag{2.68}$$

under the constraints of total time c_i available for each i in the form

$$\Sigma_j \, t_{ij} \, c_{ij} \ \leq c_i \tag{2.69}$$

Concentrating in this exposition on the case where (2.69) is an equality, we can solve the above problem for x_{ij} as

$$x_{ij} = x_{ji} \, [(b_{ij} + b_{ji}) / (2b_{ii})]^2 \tag{2.70}$$

It can be shown that each make equal contributions *if and only if* their scientific 'worths' are equal, that is $b_{ii} = b_{jj}$. In Beckmann (1994), several other interesting properties of the solution are demonstrated. For instance, i and j will not collaborate if the average interaction benefit per unit time spent $[(b_{ij} + b_{ji}) / 2 \, t_{ij}]$ is less than that $\surd(b_{ii} \, b_{jj})$ of working alone.

In the above elaboration, it is considered that the effort for working alone is independent of the bilateral interactions, just being constrained by the available time budget and acting rather like a residual. In Roy and Kobayashi (1996), we went a step further. We considered that part of the time working alone was jointly 'transformed' into output via the interaction with each collaborator. If i decides to collaborate with L individuals l and x_{iil} is taken as the time i works alone inspired by the face-to-face contacts with l, we can then define the production function f_i as

$$f_i = \Sigma_l \ \surd(x_{il} \, x_{li} \, x_{iil}) \tag{2.71}$$

where $\Sigma_l \, x_{iil}$ equals the total effort x_{ii} working alone. Of course, we could add an extra component of private work not directly attributable to any one collaborator.

An important aspect of face-to-face contacts, identified by Kobayashi and developed in Kobayashi *et al.* (1998) is the property of *mutuality*. In other words, a bilateral contact will only be realised if *both parties* perceive that they will make a net gain. This property distinguishes contacts from conventional choice processes, where a choice is made autonomously by each individual. We may note that household decision processes have some common properties with contact processes, where choices acceptable to all household members are sought.

Solutions to Numerical Exercises

Exercise 2.1

(i) $p_{01} = 21000$, $p_{02} = 16000$, $r' = 20$, $r_m = 32$, $S_1 = 251330$, $S_2 = 490090$
(ii) $v_1 = 409.23$, $v_2 = 326.08$
(iii) $S_1 = 280900$, $S_2 = 613210$. This yields divergence with a negative price v_2
(iv) The inverse supply functions come out as:

$$v_2 = z_2 + t_2 [\sqrt{(S_1 Q_1 + S_2 Q_2)}] / \sqrt{\pi}$$

$$v_1 = z_1 + \{ Q_2 t_2 [\sqrt{(S_1 Q_1 + S_2 Q_2)}] + [Q_1 t_1 - Q_2 t_2] [\sqrt{(S_1 Q_1)}] \} / [Q_1 \sqrt{\pi}]$$

(v) The converged results after 3 iterations are $v_1 = 400.15$, $v_2 = 320.08$

Exercise 2.2

(i) At this position, all stores have an equal market share of *0.25*. The social optimum, defined as that yielding the minimum average distance travelled, is also at this point, with the minimum distance equal to *0.125*.
(ii) By this move, the facility suddenly improves it's market share by 50%. This implies that the equilibrium is unstable.
(iii) However, in this case, the moved facility has a vanishingly small increase of its market share. With the market shares of all facilities being *0.25*, we have here the only stable competitive equilibrium. The minimum average distance is *0.25*.

3 Spatial Interaction Modelling Embracing Microeconomics

3.1 Evaluation of Relevant Previous Efforts

3.1.1 Re-Interpretation of a Retail Equilibrium Model

Which model framework in the spatial interaction tradition seems to have been the first that explicitly applied equilibrium concepts? Which paper most clearly points the way towards the eventual objective of this book – the enrichment of spatial interaction models by concepts from microeconomics? It is clearly the equilibrium model of Harris and Wilson (1978), introduced in Chap. 1. The objective here, guided by the classical applications of microeconomics in space summarised in Chap. 2, is to re-interpret the Harris/Wilson equilibrium framework in terms of short run and long run equilibria in economics.

Consider firstly short run consumer demand, given a configuration of retail centres j each with floorspace W_j, expressed via the Huff model introduced in Huff (1964) in the form

$$S_{ij} = O_i \; W_j \, exp - \beta t_{ij} / [\Sigma_j \, W_j \, exp - \beta t_{ij}] \tag{3.1}$$

where O_i represents the given number of shopping trips emerging from zone i, t_{ij} the travel time between zone i and centre j and β a gravity impedance parameter. We firstly note that relation (3.1) is not in the form of a demand model from economics as (i) it contains no goods prices and (ii) the total number of trips O_i, as a proxy for the total retail expenditure, is given in advance. This more truncated model structure also implies that a companion short run supply model for retail goods and services is superfluous here. Thus, the *short run* is denoted entirely via relation (3.1).

In the *long run*, it is considered that the retail capacities W_j are allowed to expand or shrink in size from the standpoint of the tenants of the retail centres in response to demand. The tenants sell goods and services and pay wages to their employees as well as rents to the owners of the centres. In this analysis, rents are given exogenously and we confine ourselves to the tenant/consumer equilibrium. In order to more fully place the Harris/Wilson model in an economic equilibrium framework, it is necessary to assume that each retail centre contains a very large number of independent tenants, each interested in profit maximization. Also, let the prices of the goods, rents and unit transaction costs be identical in each centre.

In this context, now consider the criterion for a spatial long run *Nash* equilibrium (Intriligator, 1971), which can only occur in a situation in which *if any individual retailer makes a unilateral move to another centre, he will not be better off!* Assuming a very large number of tenants in each centre, this state can only be realised if the final distribution of retailing floorspace W_j is such that the profits per unit floorspace are *identical* in each centre. If c is a constant, this implies the condition

$$(\Sigma_i S_{ij}) / W_j = c \tag{3.2}$$

where $(\Sigma_i S_{ij})$, with constant prices and unit costs, represents the total profits accruing to each centre. If it is also assumed that the total amount of floorspace to be allocated is W, and given the total number of trips $S = \Sigma_i O_i$, then by summation of the transposed (3.2) over j, we evaluate the constant c as (S / W).

Now, if the short run model (3.1) is *embedded* for S_{ij} in (3.2), we obtain a set of non-linear simultaneous equations for the long run equilibrium floorspace values. This ensures that the veracity of the short run model is retained within the long run adjustment process. In the Harris and Wilson paper, (3.2) was solved for several cases, including those where the floorspace W_j was raised to a power α. In the linear case above, the solution is obtained via simple successive substitution. With the long run adjustment of floorspace being a *slow process,* Harris and Wilson assumed a *logistic dynamic* path towards long run equilibrium. In this analysis, they uncovered a fundamental *bifurcation* between a more spatially centralised retailing system and a more spatially dispersed system, depending on the values of the parameters α and β. These insights propelled the paper to the forefront of current thinking on spatial dynamics.

From the above, it can be appreciated that the Harris/Wilson model framework has a clear interpretation in terms of short run and long run spatial economics. At the same time, it is limited in this respect by the absence of prices for the goods and services, the lack of a corresponding short run model for the supply of retailing goods and services and the assumption of inelastic demand. In Chap. 4, these restrictions are lifted.

3.1.2 Dispersed Spatial Price Equilibrium

Classical spatial price equilibrium (SPE) models, described in Sect. 2.2.2, have been successfully applied by Takayama and others to the shipment of bulk commodities, such as mineral ores and food grains. As mentioned by Harker (1988), 'The SPE model, however, is flawed in the sense that it often performs poorly in practice when it is rigidly applied, that is, when no calibrations of the model to a sample of interregional flow patterns and regional prices are made prior to any policy analysis'. As the latter calibration process is not an intrinsic part of the SPE model itself, the process can be rather ad hoc. Especially when one is modelling trade in higher value differentiated goods, the absence of cross-hauling (that is, goods being traded in both directions between regional pairs) in the classical SPE model and the high proportion of zero flows are unrealistic – the model just assumes that trade flows occur from any exporting region to just one importing region, the one yielding the greatest return net of transport costs. As demonstrated by Munroe and Hewings (1999) in relation to the new trade theory, this is no mere academic point – advanced economies with highly differentiated goods are characterised by an increasing proportion of intraindustry trade reflected in increased cross-hauling.

The first significant paper to take up the challenge of developing a more realistic spatial price equilibrium model framework was Batten and Johansson (1985), who coined the term *dispersed spatial price equilibrium* (DSPE) for their model. It not only allowed cross-hauling and exports to more than one region, but also used the historical flow patterns to allow implicitly for rigidities and affinities in trade relationships. The paper was then followed by that of Harker (1988), where the exchange or trade mechanism was depicted by a gravity model. As this is somewhat closer in spirit to the theme of this book, it will be described below. In order to represent the ideas with maximum clarity, two simplifying assumptions will be made with respect to Harker's formulation (i) only one route is assumed to exist between each exporting region and each importing region and (ii) all regions are both exporters and importers and no pure *transhipment* nodes exist.

Starting by defining symbols, let:

T_{ij} be the flow from region i to region j

A_i, B_j be pre-calibrated origin and destination 'push' and 'pull' factors

S_i be the number of trips (supply) leaving region i

D_j be the number of trips (demand) arriving at destination j

c_{ij} be the unit transport cost between regions i and j

γ be a pre-calibrated parameter

$\psi_i(\boldsymbol{S})$ be the exogenous supply price function in region i, potentially as a function of the complete supply vector $\boldsymbol{S}$

$\theta_j(\boldsymbol{D})$ be the exogenous demand price function in region j potentially as a function of the complete demand vector $\boldsymbol{D}$ and

V_{ij} be the unit shipper profit from region i to region j.

Let the supply and demand behaviour at the origin and destination regions, expressed via the exogenous supply and demand price functions, be *mediated* by shipper behaviour expressed via calibration of a special version of the doubly-constrained gravity model of Wilson (1967) which is calibrated to the most recent set of flows. Rather than applying a transport cost constraint, as illustrated by Wilson for personal travel, a shipper profit constraint is applied in terms of the unit profits

$$V_{ij} = \theta_j(\boldsymbol{D}) - \psi_i(\boldsymbol{S}) - c_{ij} \tag{3.3}$$

When the model is calibrated, the parameter γ is obtained, as well as 'balancing factors' A_i and B_j, which become exogenous parameters when the model is used for projection, as formalised in Lesse (1982) and applied by Roy (1985). This yields shipper behaviour expressed as

$$T_{ij} = A_i B_j S_i D_j \exp\gamma[\theta_j(\mathbf{D}) - \psi_i(\mathbf{S}) - c_{ij}] \tag{3.4}$$

together with the flow consistency conditions

$$\Sigma_j T_{ij} = S_i \qquad ; \qquad \Sigma_i T_{ij} = D_j \tag{3.5}$$

which connect the endogenous flows with the endogenous regional supply and demand. Relations (3.4) and (3.5) define Harker's DSPE model. We note, that in contrast to the SPE model, non-zero flows exist between all regions.

The model is implemented by substitution of (3.4) into (3.5) and solving these equations simultaneously for the unknown supply and demand quantities S_i and D_j. Because of the non-separable supply and demand price functions within the exponential term, the solution must usually be performed iteratively. Harker (1988) gives conditions for existence and uniqueness of the solutions, which are not overly restrictive.

The one disadvantage of Harker's approach is that the supply and demand functions must be provided exogenously, with the probabilistic part of the model confined to shipper behaviour. In Roy (1992) and Roy and Johansson (1993), we overcome these limitations, defining *dispersed* supply or demand functions endogenously, depending on whether the exporters or importers pay the transport costs. In Chap. 7, further models of this type are developed.

3.2 Comprehensive Specification of a Unifying Framework

This section is the kernel of the book. With a mind lubricated by the heroic efforts of the past, as discussed in Chaps. 1 and 2, you, the reader, are now in the position to anticipate the drawing together of the diverse threads from these past accomplishments to gain a fuller understanding of this evolving field. This is illustrated for not only commodity flow models, but for retailing models and journey-to-work models.

3.2.1 Entropy in a Decision-Theoretic Context

Traditionally, most entropy models, as in Wilson (1967, 1970), describe a process or a phenomenon. For instance, the key model in Wilson (1967) depicts the process of commuting. However, if we are going to achieve a greater consistency with economic theory in this class of model, we must first attempt to identify the *individual agents* who operate in the system being examined. Following this, we then characterise the behaviour of each set of agents. For the simple commuting example, we attempt to model the behaviour of segments of commuters identified by their zone of residence *travelling* to various categories of jobs in the same or other zones. However, in a wider sense, we may be interested in modelling the longer term *choice of job* by workers, which implies an *expectation* of the required commuting effort. In this case, the model becomes one of *spatial demand* for jobs, which should be supported by *spatial supply* models from the point of view of the firms (Roy, 1990). Such models must then recognise the possibility of unemployment and unfilled vacancies. In addition, in contrast to travel models, they should be based on the *stated* values of travel times *expected* by the worker when making the choice of job (Roy, 1989). Such times can only be obtained by a *stated preference survey.* In addition to pure demand or supply agents, the economy is characterised by an increasing proportion of *transformative* agents, who demand inputs and supply more sophisticated transformed outputs, as described, for example, in Roy (1994).

The individual agents and their individual choices or options then become the basis for specification of the microstates for each problem, with *individuality* being expressed as *distinguishability* of the individuals and/or choices which define these microstates (Roy and Lesse, 1981). **In this endeavour, we ensure that the *microlevel* for the entropy formulation coincides precisely with the *agent level* in microeconomics.** Though the analysis itself is carried out at a more aggregate level, such as for sets of demand firms and sets of supply firms, the entropy microstate *objective* denotes the *choice freedom* available within each group at the fundamental level of individual agents.

3.2.2 Different Entropies for Different Decision Contexts

The decisions being modelled involve individual agents making individual choices from the sets currently available. As sketched in Sect. 1.2.5, different definitions of distinguishability of agents and their choices, as well as different characteristics of the choice sets, lead to different entropy types.

For the first illustration, consider the conventional commuting model of Wilson (1967) in the form of an equilibrium job choice process. Let O_i workers from zones i choose jobs D_j in job zones j in unknown groups T_{ij}, where the total number of workers equals the total number of jobs. As both the individual workers in a residential zone and the specific jobs in an employment zone vary with respect to the respective influence of commuting requirements on job choice, the key purpose of the model, they are both treated as distinguishable in the microstate definitions. This distinguishability has links to the assumption of Armington (1969) in trade economics. Thus, the number of microstates M_I becomes the number of ways that O_i distinguishable commuters may be allocated to job zones j times the numbers of ways they may be allocated to distinguishable jobs within the set D_j, yielding

$$M_I = \{\pi_i O_i! / [\pi_j T_{ij}!]\} . \{\pi_j D_j!\} \quad (3.6)$$

where π represents the product sign and *!* the factorial operation. When the natural log of this expression is taken and the Stirling approximation applied, the entropy S_1 comes out as

$$S_1 = -\Sigma_{ij} T_{ij} [\log T_{ij} - 1] \tag{3.7}$$

in which the known values O_i and D_j are superfluous in the objective. Now, how would this expression change if there were unemployment, that is, the number of employed workers O_i were less than the number N_i in the workforce in each residential zone i, as given in Roy (1990)? Then, the number of microstates M_2 would be given as the number of ways that the total N_i distinguishable workers can be allocated into unknown groups of unemployed workers $(N_i - \Sigma_j T_{ij})$ times the number of ways the distinguishable employed workers are allocated in groups T_{ij} from each residential zone i into each job zone j times the number of ways they are allocated to distinguishable jobs among the total of D_j available, yielding the new expression

$$M_2 = \{\pi_i N_i! / [(N_i - \Sigma_j T_{ij}!)] [\pi_j T_{ij}!]\} . \{\pi_j D_j!\} \tag{3.8}$$

As before, this leads to the entropy S_2 in the form

$$S_2 = -\Sigma_i (N_i - \Sigma_j T_{ij}) [\log (N_i - \Sigma_j T_{ij}) - 1] - \Sigma_{ij} T_{ij} \{\log T_{ij} - 1] \tag{3.9}$$

Under the implied assumption of distinguishability of the job applicants, the model based on the above objective function ensures that, so long as there is unemployment in the total system, there will be at least some amount of unemployment in *each* residential zone. This is a much more plausible hypothesis than having unemployment concentrated at isolated points within a region.

The above models assume that a job can only be occupied by one worker at a time. However, in other situations, such single choice trips are replaced by multiple choice trips or the choices overlap. For example, in a large shopping centre, customers can potentially purchase different categories of goods or consume different categories of personal services. In order to cope with this product variety, let's assume that a centre j is subdivided into W_j so-called *elemental shops*. Consider firstly S_j distinguishable customers each choosing one among the W_j distinguishable elemental shops, where different customers can choose to shop at the same elemental shop. Then, for all centres j, the number of microstates M_3 comes out as

$$M_3 = \pi_j W_j^{S_j} \tag{3.10}$$

In practice, two countervailing influences occur. Firstly, most customers choose more than one elemental shop during a visit. Secondly, the complete number of elemental shops may not be in the habitual choice set of any chosen customer. This leads to a scaling effect in (3.10), which in log form yields the entropy S_3 as

$$S_3 = \Sigma_j S_j \log W_j^{\alpha} \tag{3.11}$$

where the customer scaling parameter α is calibrated via the method of maximum likelihood, simultaneously with calibration of the other model parameters as Lagrange multipliers on the model constraints (Roy, 1983).

It is important to emphasise that microstates with regard to distinguishable or heterogenous choices are not relevant when the agent has no choice freedom in the situation being modelled. For instance, though the beds in a public hospital are different to each other (some may have better views or be more centrally located), the patients usually have no freedom of choice of bed - they are merely allocated to the first available bed. Thus, there are no microstates defined with respect to individual beds - the patient is allocated to the hospital as a whole. This converts a potential 2-level choice to a simple single level choice. Similar arguments apply in any situation where an individual is externally allocated a specific item within an aggregate set of items.

3.2.3 Fundamental Role of Constraints in Modelling Supply and Demand

As illustrated in Chap. 1, 'information' is added to the model in the form of constraints based on observations at a chosen base period. The model is 'calibrated' such that the modelled results conform to the base period observations for each of the constraints. For commodity flow or personal travel models, some of these constraints refer to the sums of the rows or the columns (total outflows and inflows). In other cases, they represent quantity constraints on capacity. Such constraints are discussed in detail in a later section. In this section, we examine the 'behavioural' constraint or constraints, which are indispensable, and specific to the system being modelled.

In the path-breaking work of Wilson (1967), the behavioural constraint used for transport models was the average trip time or cost over the entire system, usually treated as a *generalised cost*, where the value of the trip time is added to the money cost of the trip. Such constraints continue to be applied in such models. Their role was strengthened by the demonstration by Evans (1973) that, in the limit, as the Lagrange multiplier on this travel cost constraint approaches infinity, the entropy solution approaches asymptotically that of the 'transportation problem' in linear programming, where the flow pattern is sought which minimises the total transport costs across the entire system. As seen in Fig. 1.1, the gravity travel cost parameter β varies between zero and infinity as the entropy reduces from its maximum value towards zero. This identification of the limiting case of the entropy problem having a transport cost **constraint** with a mathematical programming problem where this constraint now becomes the **objective** is the key step in providing a clear behavioural basis for entropy models, in which such classical deterministic maximization or minimization problems emerge as special cases of our probabilistic entropy formulations.

Smith (1990) was the first to fully appreciate the significance of the behavioural constraint in entropy or most probable state models, which he described as an 'information' constraint. Also, he demonstrated (see Sect. 1.3.5) the fundamental asymptotic property of this class of model. We re-quote some of his statements *"The choice of the information constraints is no longer arbitrary.... Indeed, these constraint values are now asserted to be the fundamental macro variables of the theory itself...Each choice of macro variables constitutes a different theory of micro behaviour, and hence this choice is of major importance."*. And, *'If theory T is true, then for all sufficiently large population realizations, the maximum entropy state is the* ***overwhelmingly most probable*** *macro state which can occur".* Finally, *"These theories allow many types of* ***strong interaction effects*** *among micro behaving units to be incorporated"*. In the absence of these latter properties, we would really be floundering! In our case, the theory of micro behaviour to be reflected in the information constraint is **microeconomic theory**. This concept is illustrated below for both supply and demand modelling.

Consider the classical spatial **supply** problem of microeconomics, where firms purchase some of their inputs from other regions, paying the required transport costs. Let x_i^{rs} be the flow of input commodity i between regions r and s and let there be n such input commodities going into the production of output commodity j in regions s (the index j is suppressed for convenience). Let $\overline{p}_s$ be the unit price for the output commodity in region s and let p_i^r be the unit cost of input commodity i produced in region r. Also, let c_i^{rs} be the unit transport cost of one unit of commodity i between regions r and s. Finally, assume that all the output firms (for commodity j) in a given region s operate with the same technology, and let the total inputs $(\Sigma_r x_i^{rs})$ of each commodity i used in region s be transformed to output Z^s according to the following concave production function, taken to be already calibrated

$$Z^s = f^s(\Sigma_r x_1^{rs},...\Sigma_r x_i^{rs},....\Sigma_r x_n^{rs}) \tag{3.12}$$

Assuming identical output firms, let the classical profit maximization problem of microeconomics relate to maximizing the total profits Π in the system in terms of the unknown input demand requirements x_i^{rs}. For completeness, we consider the F^s firms in region s paying (exogenous) rents r_s. As this short run problem does not consider re-location, the Spatial Impossibility Theorem of Fujita and Thisse (2002) does not apply.[1] with Π expressed as

$$\Pi = \Sigma_s \overline{p}_s Z^s - \Sigma_{irs} x_i^{rs} (p_i^r + c_i^{rs}) - F^s r_s \tag{3.13}$$

where the production function Z^s is inserted from (3.12). In the classical deterministic input demand problem, expression (3.13) is maximised in terms of

[1] Of course, their results must be accounted for in any long run locational extensions. A stable short run equilibrium is of little comfort when the long run milieu in which it is nested is unstable!

the input flows x_i^{rs}. Once the profit-maximizing **input demand functions** x_i^{rs} are determined, they can be substituted into (3.12) to produce the corresponding **supply functions.**

There are many reasons why the above classical solution is not realised in practice. These include imperfect information available to the firms, differences in technology between the so-called identical firms and differences in strategic objectives. In fact, if for a certain base period we have enough commodity flow data to evaluate the actual realised profits Π^0 by substitution of the *observed* flows into (3.13), the resulting total profits can never be greater than the results of the classical deterministic solution, and will often be considerably less. Thus, if we are interested in projecting the state of the spatial supply system at a future point in time, why use the deterministic objective (3.13), which is usually unrealistic from the start. Instead, we formulate and *fit* a model to reproduce the *observed* total profits Π^0, which at the same time has the important asymptotic property that it converges in the limit to the classical solution of (3.13). **This correspondence is a major theme of this book.** Let us now proceed.

Consider the known inputs (stock) X_i^{r0} of each commodity i in each producing region r, as well as the unknown usage $X_i^s = \Sigma_r x_i^{rs}$ of each commodity i for the output of the chosen commodity j in each producing region s. If we want to allow for different levels of productivity of the firms in different regions, we would not identify the firms with individual shipments, as above, but would look at the receiving firms F^s absorbing the inputs. Thus, we consider the number of ways M_a that the X_i^{r0} distinct observed shipments from region r can be allocated in groups x_i^{rs} to regions s and the number of ways the $X_i^s = \Sigma_r x_i^{rs}$ shipments arriving at region s can be arbitrarily allocated to the F^s distinct receiver firms there, yielding

$$M_a = \pi_{ir} X_i^{r0}! / [\pi_s x_i^{rs}!] \quad . \quad \pi_{is} F^{s\,X_i^s} \tag{3.14}$$

Now, assuming that we are going to reproduce the observed input flows X_i^{r0} of each commodity i out of each input region r, which the firms at regions s compete for, an extra calibration feature not available to the classical deterministic model, the following row sum constraints should be applied

$$\Sigma_s x_i^{rs} = X_i^{r0} \tag{3.15}$$

If the natural log is taken of (3.13), the Stirling approximation applied and constant terms omitted, we obtain the flow entropy S_a in the form

$$S_a = - \Sigma_{irs} x_i^{rs} [\log (x_i^{rs} / F^s) - 1] \tag{3.16}$$

Now providing the objective (3.13) of the deterministic problem as a constraint, the key behavioural constraint becomes

$$\Sigma_s \bar{p}_s Z^s - \Sigma_{irs} x_i^{rs} (p_i^r + c_i^{rs}) - F^s r_s = \Pi^0 \tag{3.17}$$

where the production function Z^s is inserted from (3.12) as before and the 'observed profits' Π^o are evaluated from base period observations. If the entropy (3.16) is maximized under the row sum constraints in (3.15) with Lagrange multipliers λ_{ir} and the key behavioural constraint with multiplier $-\alpha$, the **probabilistic input demand functions**, after eliminating λ_{ir}, become

$$x_i^{rs} = X_i^{r0} F^s \exp \alpha [\bar{p}^s \{\partial Z^s(....)/\partial x_i^{rs}\} - (p_i^r + c_i^{rs})] / \Sigma_s \{ F^s \exp \alpha [\bar{p}^s \{\partial Z^s(....)/\partial x_i^{rs}\} - (p_i^r + c_i^{rs})]\} \quad (3.18)$$

The solution of this problem may be obtained by successive substitution so long as the production function has its usual concavity properties. When (3.18) is inserted into the production function (3.12), we obtain the **probabilistic (expected) supply function.**

In order for the reader to better appreciate the comparison between the two approaches, the production function Z^s is specialised to the following common Cobb-Douglas form, with $(\Sigma_s x_i^{rs})$ denoted as X_i^s, yielding

$$Z^s = b_0^s \; X_1^{s\,b_1^s} \dots X_i^{s\,b_i^s} \dots X_n^{s\,b_n^s} \quad (3.19)$$

in which a typical derivative $\partial Z^s / \partial x_i^{rs}$ reduces to

$$\partial Z^s / \partial x_i^{rs} = b_i^s Z^s / x_i^{rs} \quad (3.20)$$

Using this result in the maximization of the classical deterministic profit function, it is easy to show that the profit maximizing input demand flows come out as

$$x_i^{rs} = b_i^s \; \bar{p}^s Z^s / (p_i^r + c_i^{rs}) \quad (3.21)$$

which, with the production function Z^s being a function of the input flows of all the commodities i, must be solved by successive approximation.

Now, coming to the entropy maximisation solution (3.18), the expression (3.20) is again substituted for the partial derivatives and the unknown Lagrange multipliers determined, together with the unknown commodity flows, via an iterative process. Using the fundamental results of Evans (1973), it can be seen that as the calibrated Lagrange multiplier α on the behavioural observed profit constraint (3.17) approaches infinity, the solution (3.18) of the probabilistic entropy problem approaches asymptotically that of its deterministic counterpart in (3.21), upon two provisos. Firstly, the deterministic model is not forced to conform with the observed row sum constraints (3.15), but merely with a single quantity constraint based on the overall flow total. Secondly, the number of firms F^s in region s, which yields an *à priori* bias for our entropy models to allocate inputs to regions s in proportion to the number of firms there, is not present in the deterministic model. This example shows the *classical deterministic solution from microeconomics to be a special case of our probabilistic entropy result.*

We now turn to the corresponding problem in **demand** theory, where consumers are taken to maximise a utility function subject to a budget constraint. As before, let the consumers of the final demand travel from their home region r to a consumption region s to buy or consume the product, absorbing the transport costs themselves. Let y_i^{rs} be the amount of commodity i consumed by residents of region r at region s. Let p_i^s be the purchase price of a unit of commodity i at region s and let c_i^{rs} be the generalised transport cost (time plus money costs) of travel between regions r and s times the trips per unit quantity of good i consumed there in each region visited, taken to be absorbed by the buyers[2]. Let a utility function U^r be defined and calibrated over the consumers of region r (assumed to be identical) in terms of the quantities $(\Sigma_s y_i^{rs})$ consumed via

$$U^r = F^r(\Sigma_s y_1^{rs}, \ldots \Sigma_s y_i^{rs}, \ldots \Sigma_s y_m^{rs}) \tag{3.22}$$

where there are m consumption commodities. If there are N^r consumers in each region r, then we must satisfy a budget/income constraint related to the average available income I^r of each consumer, yielding

$$\Sigma_{is}\, y_i^{rs}(p_i^s + c_i^{rs}) = N^r I^r \tag{3.23}$$

This simplified spatial version of the classical utility maximization model can be readily solved, so long as the utility function has the required concavity properties. In line with the supply problem, the total system utility $\Sigma_r U^r$ is maximized.

As for the realised profits of the supply problem, the maximum realised (optimum) utility U^r emerging from the classical deterministic solution will exceed that U^{r0} computed from the observed flows y_i^{rs0}. Thus, we investigate an entropy approach to cope with this divergence. In defining the number of microstates, we consider the number of ways M_b that the observed number $Y_i^{s0} = \Sigma_r y_i^{r0s}$ of distinct units of each commodity i ordered at each region s may be allocated to consumer regions r in groups y_i^{rs} times the number of ways the total of distinct commodity orders $Y_i^r = \Sigma_s y_i^{rs}$ made from region r may be arbitrarily allocated to each of the N^r consumers there, yielding, as the mirror-image of the supply entropy in (3.14), the result

$$M_b = \pi_{is}\, Y_i^{s0}! / (\pi_r\, y_i^{rs}!) \;\; . \;\; \pi_{ir}\, N^{r\,Y_i^r} \tag{3.24}$$

If the log of (3.24) is taken and the Stirling approximation applied, then the entropy objective S_b comes out as

$$S_b = -\Sigma_{irs}\, y_i^{rs}\,[\log(y_i^{rs}/N^r) - 1] \tag{3.25}$$

---

[2] This complex definition is an approximation, resulting from the lack of a one-to-one correspondence between retail travel costs and quantities purchased (see Chap. 4 for a resolution). Note, a direct correspondence does exist for commodity shipments.

The column sum constraints are applied to the orders from regions s, yielding

$$\Sigma_r y_i^{rs} = Y_i^{s0} \tag{3.26}$$

Now, if we have the total utility U^0 based on observed orders in all demand regions r, the following behavioural constraint is applied

$$\Sigma_r F^r (\Sigma_s y_1^{rs}, \dots \Sigma_s y_i^{rs}, \dots \Sigma_s y_m^{rs}) = U^0 \tag{3.27}$$

If the entropy (3.25) is maximized under the constraints (3.23) with multipliers ϕ^r, (3.27) with multiplier β and (3.26), we obtain

$$y_i^{rs} = Y_i^{s0} N^r \exp\{\beta [\partial F^r(\dots)/\partial y_i^{rs}] - \phi^r (p_i^s + c_i^{rs})\} \;/\; \Sigma_r \{N^r \exp\{\beta [\partial F^r(\dots)/\partial y_i^{rs}] - \phi^r (p_i^s + c_i^{rs})\}\} \tag{3.28}$$

Except for the budget constraint (which is satisfied as an equality in this class of problem), the solution is a mirror image of (3.18) for the supply model. However, there are two further points. First, the definition of the travel costs c_i^{rs} for the demand model are only transparent if the customers make orders which are delivered to them. However, if they travel to shop, such as within a large urban region, we do not have a one-to-one correspondence between trip costs and quantities purchased – they may spend more or less than usual on any one trip. A method for handling this difficulty is presented in the next chapter. Second, as the externally estimated parameters of the production functions for the supply model only denote the input quantities to output quantity transformation, they do not relate *directly* to the ultimate profit objective. On the other hand, the externally estimated parameters of the utility function for the demand model represent the objective itself. Thus, our use of the entropy in the supply model based on 'observed' profits may more significantly differ from the deterministic solution than is the case for the demand model, where the calibration of the utility parameters will already be based on some sort of base period observations.

Again we illustrate the general procedure by choosing a form of utility function, this time assumed to be linear, which is quite usual in many applications. Thus, the total U is written as

$$U = \Sigma_r a_0^r + a_1^r Y_1^r \dots + a_i^r Y_i^r \dots + a_m^r Y_m^r \tag{3.29}$$

The maximization of (3.29) should be performed under a budget constraint for each region r, as implied above in (3.28). As the problem is a classical linear program, the solution is expressed numerically. For the entropy problem, the solution in (3.28) with a linear utility function comes out simply as

$$y_i^{rs} = Y_i^{s0} N^r \exp[\beta a_i^r - \phi^r (p_i^s + c_i^{rs})] / \{\Sigma_r N^r \exp[\beta a_i^r - \phi^r (p_i^s + c_i^{rs})]\} \tag{3.30}$$

which has a similar form to a logit model. Again, in the limit, and omitting the column sum constraints (3.26) and the number of consumers bias term N^r, it is easy to demonstrate that the solution of (3.30) approaches asymptotically that of the linear program for the deterministic problem.

Thus, for both the supply and demand problems, we have demonstrated that **the classical solutions from microeconomics, where assumed identical producers or consumers are aggregated by region, are special limiting cases of our entropy solution.** We can go further. As illustrated in the next chapter, we can introduce *time budget* constraints from the perspective of **the theory of time geography** into the above demand model, which then conforms with both microeconomics and time geography principles. On the supply side, we can replace the perfect competition assumption by *imperfect competition* considerations. As for the classical solutions, this works elegantly when the demand functions are analytically invertible, which is demonstrable for our general demand model in (3.28).

Of course, it would be possible to introduce other *smoothing techniques* in addition to entropy. However, entropy formulations are ideal where a microlevel is identifiable, but where analysis for various reasons is carried out at a more macrolevel. This is precisely the case for the above formulations and for much of urban and regional analysis.

3.2.4 Calibration of Open Models Adding Information via Constraints

In the first instance, we must regard model **calibration** merely as solution of the constrained entropy maximization problem for both the unknown flows and the unknown Lagrange multipliers on constraints using data at a certain base period. In attempting to give probabilistic versions of the *open* models of economics, we are faced with an apparent contradiction. For our models to be information-rich and thus robust, *information* needs to be added to the model system during *calibration* in the form of constraints. On the other hand, such constraints would appear to *close up* the model when it is being used for projection. This conundrum is handled by using a formalism introduced by Lesse (1982). This formalism was first used to open up spatial interaction models in a paper by the author (Roy, 1983). Despite the elapsed time, the power of the approach seems not to have been generally appreciated. These ideas are developed below, as applied to the Huff retail model (see Sect. 1.1.2), where (i) we calibrate the model to the destination inflows rather than using a uniform power of floorspace as the attractor and (ii) we remove the classical origin constraints by allowing trips to be generated arbitrarily by the households. Improvements (i) and (ii) can either be applied together or separately. The first adjustment, described in Roy (1983), was motivated by an actual application of the Huff model to the city of Cairns in the tropical North of Australia. State the Huff model in its usual form

$$S_{ij} = O_i W_j^{\alpha} exp - \beta t_{ij} / (\Sigma_j W_j^{\alpha} exp - \beta c_{ij}) \qquad (3.31)$$

where S_{ij} represents the number of shopping trips between residential zone i and retail centre j, O_i represents the number of shopping trips emanating from zone i,

W_j is the floorspace of centre j, t_{ij} is the travel time between zone i and centre j, α is a floorspace attractiveness parameter and β a gravity impedance parameter, both of which are calibrated as part of the entropy maximization process. If we look at the number of trips $S_j = \Sigma_i S_{ij}$ attracted to any centre j, we see that they are jointly proportional to accessibility $(\Sigma_i O_i \exp - \beta t_{ij})$ and the power W_j^{α} of the floorspace. However, in Cairns, the CBD was the largest and most accessible retail centre, but was much less attractive to customers than the model would imply. This was because (a) the CBD at that time was rather run down, (b) shoppers had to go out of an air-conditioned shop into the tropical heat of the street to enter additional shops and (iii) the new retail centres on the outskirts of the city were fully undercover and air-conditioned. As a result, the calibration process of the normal Huff model (3.31) diverged! We were faced with the challenge of coming up with a revised version of the model allowing the behaviour 'on the ground' to be represented. The main limitation of the existing model was that the centre attractiveness (W_j^{α}) is a *uniform* power α of the floorspace, rather than being different for centres of different perceived 'quality'. The first step is to *calibrate* the retail model as *doubly-constrained*, by applying the regular origin and travel time constraints, plus constraints on the number of shoppers S_j at centres j via

$$\Sigma_j S_{ij} = O_i \quad ; \quad \Sigma_{ij} S_{ij} t_{ij} = T \quad ; \quad \Sigma_i S_{ij} = S_j \tag{3.32}$$

where T is the total travel time in the system. In the past, such destination constraints were only considered relevant for commuting models (Wilson, 1967). If all these constraints are included, and the linear floorspace is extracted from the destination Lagrange multipliers, we have the revised model

$$S_{ij} = O_i A_i W_j B_j \exp - \beta t_{ij} \tag{3.33}$$

in which $A_i = \exp - \lambda_i$ and $B_j = \exp - \eta_j$, the negative exponentials of the respective origin and destination constraint Lagrange multipliers λ_i and η_j, and β is the Lagrange multiplier on the travel time constraint, we obtain the following recursive relations (as in the usual doubly-constrained model) via

$$A_i = (\Sigma_j W_j B_j \exp - \beta t_{ij})^{-1} \quad ; \quad B_j = (S_j/W_j)(\Sigma_i O_i A_i \exp - \beta t_{ij})^{-1} \tag{3.34}$$

In the *calibration* process, equations (3.34) are solved iteratively using successive substitution, *simultaneously* with insertion of (3.33) into the middle equation of (3.32) and solution of the resulting non-linear equation for β using linear extrapolation. We note that the non-linear constraint equations may alternatively be solved using **Newton-Raphson** iteration. However, this more general approach involves a matrix inversion and may converge more slowly. Now, in the commuting doubly-constrained model (Wilson, 1967) it is assumed that we can *exogenously* estimate the number of jobs at each employment zone j. On the other hand, the number of shoppers S_j visiting each centre j is an expected *endogenous* output from a retail model, and it is inappropriate to apply constraints (3.32) when making future projections. How do we cope with this difficulty? As with the gravity *Lagrange multiplier* β, we also transform the destination balancing factors B_j into *parameters* for future projection. This is handled by the formalism of Lesse (1982), where we apply to the entropy Lagrangian Z^0 at calibration

$$Z^0 = - \Sigma_{ij} S_{ij} [\log (S_{ij}/ W_j) - 1] + \Sigma_i \lambda_i (O_i - \Sigma_j S_{ij}) + \Sigma_j \eta_j (S_j - \Sigma_i S_{ij}) + \beta (T - \Sigma_{ij} S_{ij} t_{ij}) \tag{3.35}$$

a series of invariant Legendre transforms. Now, if we want both β and η_j to be treated as known parameters in (3.35) rather than as unknown Lagrange multipliers, and their right-hand sides S_j and T as unknown output in (3.35) rather than as known input, we make the following invariant Legendre transforms to the original Lagrangian Z^0, in the form

$$Z' = Z^0 - T \, \partial Z/\partial T - \Sigma_j S_j \, \partial Z/\partial S_j \tag{3.36}$$

in which the transformed Lagrangian emerges as

$$Z' = - \Sigma_{ij} S_{ij} [\log (S_{ij}/ W_j) + \beta t_{ij} + \eta_j - 1] + \Sigma_i \lambda_i (O_i - \Sigma_j S_{ij}) \tag{3.37}$$

This process removes from the Lagrangian the right hand sides of the constraints which are being targeted, that is, T and S_j in (3.32), with their associated Lagrange multipliers now being interpreted as the known values β and η_j, replacing the known values T and S_j of the original Lagrangian Z^0 in (3.35).

Now, if the transformed Lagrangian Z' is maximized in terms of S_{ij} *given* the values of the parameters β and η_j, the theory of Legendre transforms, enunciated in Lesse (1982), indicates that not only are the flows S_{ij} identical to those in (3.33) from the untransformed objective Z in (3.35), but that when these S_{ij} values are substituted into the original constraints on T and S_j, they are endogenously satisfied. The transformed model emerging from maximization of (3.37) is

$$S_{ij} = O_i B_j W_j \exp - \beta t_{ij} /)\Sigma_j B_j W_j \exp - \beta t_{ij}) \tag{3.38}$$

after elimination of the unknown Lagrange multipliers λ_i on the origin constraints.

At first sight, the above may appear to be a mere intellectual exercise. This is assuredly not the case. Having calibrated the parameters β and η_j on the base period data in the initial 'calibration' problem (3.33)-(3.34), the transformed problem (3.38) allows us to project the revised flows at a future period under potentially revised trip origins O_i, retail floorspace distributions W_j and travel times t_{ij}. At the same time, the model is enriched, compared with the Huff model in (3.31), by the destination balancing factors B_j, obtained by inducing the model at the base period to perfectly reproduce the number of shoppers S_j at each centre. In the Cairns example, the model was thus calibrated to the 'less than expected' number of shoppers visiting the CBD.

There is one further issue to consider. In the doubly-constrained solution (3.33)-(3.34), the balancing factors A_i and B_j are only known *relatively*, with one degree of indeterminacy, as pointed out by Wilson (1967). For cosmetic reasons, we carry out a simple scaling of these factors, setting the value of the median destination balancing factor B_j to unity. Furthermore, we want to find a way to *update* the destination balancing factors B_j in the projection problem (3.38) when certain key characteristics of the centres themselves change. Guided by the power form of the floorspace in the Huff model (3.31), and noting our prior linear extraction of floorspace from the centre attractiveness B_j, we can calibrate the B_j's to these other key characteristics b_{jk} and perform the following log-linear regression task

$$\log B_j = (\alpha' - 1) \log W_j + \Sigma_k \alpha_k b_{jk} \qquad \forall j \tag{3.39}$$

evaluating the parameters α' and α_k, where the floorspace power α' is clearly different to that α for the original Huff model (3.31). Typical k characteristics could include estimates of parking times at each facility, dummy variables such as the presence or otherwise of a fully air-conditioned environment or the presence or otherwise of undercover parking. Thus, in the projection model (3.38), we can change not only the trip origins, floorspace values and trip times, but, via (3.39), also the floorspace attractiveness terms B_j resulting from changes in the chosen centre characteristics.

Finally, we come to one of the central issues to be addressed to enable spatial interaction models to share an essential property of microeconomic models, that is, the endogenous response of 'quantities' (e.g. trips generated) to 'prices' (e.g. gravity impedance). We note that for the above formulations, the number of retail trips O_i generated out of each residential zone must be supplied *exogenously* to the projection model (3.38). However, trip generation clearly responds to changes in the number of households in each zone, as well as to changing accessibility to retailing (that is, changed travel times and changed provision of floorspace). We cope with this challenge by treating the number of households H_i in each zone as the generators of the O_i trips emanating from that zone. Thus, the number of microstates M_a reflects the number of ways that O_i distinguishable trips may be allocated arbitrarily (that is, some households may make more trips than others) to the H_i distinguishable households times the number of ways that the $O_i = \Sigma_j S_{ij}$ trips emanating from zone i may be allocated to retail centres j in groups S_{ij}, yielding the interesting result

$$M_a = \pi_i H_i^{\Sigma_j S_{ij}} \cdot \pi_i (\Sigma_j S_{ij})! / [\pi_j S_{ij}!] \tag{3.40}$$

Upon taking the natural log of (3.40), applying the Stirling approximation in the usual way, realising the number of trips O_i generated from each zone is going to be supplied *exogenously* for calibration and applying the W_j^{α} floorspace bias term (as justified via microstates in Chap. 1), the entropy S_a emerges as

$$S_a = \Sigma_{ij} S_{ij} [\log (S_{ij} / H_i W_j^{\alpha}) - 1] \tag{3.41}$$

which, as required, is now *driven* by the numbers of households H_i. If (3.41) is maximized under the first two of the constraints (3.32), we obtain the original Huff model of (3.31)! Defining the Lagrangian Z_a of this calibration problem as

$$Z_a = - \Sigma_{ij} S_{ij} [\log (S_{ij} / H_i W_j^{\alpha}) - 1] + \Sigma_i \lambda_i (O_i - \Sigma_j S_{ij}) + \beta(T - \Sigma_{ij} S_{ij} t_{ij}) \tag{3.42}$$

we can now look at the projection Lagrangian Z_a', where we would like to treat both λ_i and β as parameters, via the following Legendre transform

$$Z_a' = Z_a - T\, \partial Z_a/\partial T - \Sigma_i O_i\, \partial Z_a/\partial O_i \tag{3.43}$$

allowing Z_a' to be expressed as

$$Z_a' = - \Sigma_{ij} S_{ij} [\log (S_{ij}/H_i W_j^{\alpha}) + \lambda_i + \beta t_{ij} - 1] \tag{3.44}$$

where the origin push factors $A_i = exp - \lambda_i$ have been evaluated in calibration via

$$A_i = (O_i/H_i) \Sigma_j W_j^{\alpha} exp - \beta t_{ij} \tag{3.45}$$

Maximization of (3.44) yields the following transformed model

$$S_{ij} = H_i A_i W_j^{\alpha} exp - \beta t_{ij} \tag{3.46}$$

Now, in the projection period, where we assume that the parameters β and λ_i remain unchanged from the calibration period, we may now write the projected flows S_{ij}' when the number of households changes to H_i', the floorspace values to $W_j'^{\alpha}$ and the travel times to t_{ij}' as

$$S_{ij}' = H_i' A_i W_j'^{\alpha} exp - \beta t_{ij}' \tag{3.47}$$

Although the Legendre transforms provide the mathematical formalism to convert Lagrange multipliers to parameters, we still should assure ourselves that their assumed constant value (at least over the short term) is plausible. The case has already been well argued for the gravity impedance parameter β. Upon examination of (3.45), we see that the assumed constancy of the origin push factors A_i implies that the *endogenous* number of trips O_i' generated in the projection period can be obtained in terms of the following ratios between projection and base period values

$$(O_i'/O_i) = (H_i'/H_i) . [(\Sigma_j W_j'^{\alpha} exp - \beta t_{ij}') / (\Sigma_j W_j^{\alpha} exp - \beta t_{ij})] \tag{3.48}$$

This has a highly plausible interpretation. It says that the ratios between the projection period and the base period trips generated from each origin zone i is jointly proportional to the ratio of the households undertaking the trips and the ratio of their accessibility to retailing from each origin zone *i*.

Such conversions of Lagrange multipliers to parameters will be used extensively in the rest of this book. They are essential in opening up the models to reproduce the properties of the classical supply and demand models of economics, as well as in allowing the addition of relevant base period information to the model to provide greater potential reliability in projection without over-constraining the model in the projection phase. Note also, that the both types of transformations described separately above could be introduced simultaneously, yielding a Huff-type model with both endogenous trip generation and destination-specific attractiveness parameters based on nominated characteristics.

3.2.5 Nesting, Clustering and Independence from Irrelevant Alternatives

The distribution of activities in physical space is very far from uniform. In fact, many activities, such as retail facilities, tend to form clusters. This also occurs in the case of personal and producer services. Most of the earlier spatial interaction models had a structure making it impossible to identify these effects. This anomaly can be best illustrated using the Huff model.

It was in the innovative work of Fotheringham (1983) that this fundamental weakness (IIA) was described most clearly. Consider for illustrative purposes a point of concentrated population around which three identical retail centres are arranged uniformly around the circumference of an outer circle (ie. each centre subtending an angle of 120 degrees). Now the classical Huff model [see eq (1.6)] would imply, that the market shares of the three centres were equal, that is {1/3, 1/3, 1/3}. But, what if the centres were shifted around the circumference such that one was isolated and the other two were opposite and clustered together at 180 degrees to the isolated centre? Well, the original Huff model would again imply that the market shares would remain unchanged at {1/3, 1/3, 1/3}!! However, if the centres were truly identical, the isolated centre and the cluster *pair* would be expected to share the market equally, leading to market shares of {1/2, 1/4, 1/4}, as depicted in Fig. 1.2. It is clear that, in order to capture these spatial dependencies in the distribution of centres, models such as the origin-constrained model are insufficient. As a result, Fotheringham (1983, 1986) developed his set of 'competing destinations' models, which could potentially pick up these dependency effects (see Sect. 1.3.5).

At about the same time, inspired by the work of Boyce *et al.* (1983), Roy (1985) and Roy and Lesse (1985) were approaching the dependency problem from another point of view. From the knowledge of the structure of individual choice models, such as *nested logit*, it became clear that both Boyce's multilevel models and the models of Fotheringham had one key unifying property, namely they were free of the IIA (Independence from Irrelevant Alternatives) restriction. Models possessing this restriction, such as the classical Huff model described above, are incapable of changing the *relative* market shares of existing facilities when a new facility is added, even when it is placed adjacent to an existing facility. Thus, they inherently cannot identify spatial dependency influences. In the following, the key elements of Roy and Lesse (1985) are outlined, where the addition of an *entropy constraint* at a higher level of aggregation than the distribution being sought is demonstrated to not only overcome the IIA restriction, but reproduce the structure of the *nested logit* model developed for individual choice analysis (Hensher and Johnson, 1980). In the second part, we develop the *nested cluster* model of Roy (1985), where the interdependent choice of a retail cluster and an individual facility within that cluster is analysed.

Consider the choice of housing in zones i given that each commuter has a job in zones j. As illustrated in Sect. 1.3.3, we use the probabilistic form of entropy to more clearly illustrate the relationship with probabilistic logit models. Thus, consider that we know the probabilities $\overline{p}_j$ that a commuter chosen at random has a job in zone j. We seek the probability p_i that a commuter chosen at random chooses a residence in zone i, as well as the conditional probability $p_{j/i}$ that a resident choosing to live in zone i has a job in zone j, where the joint probability p_{ij} of both choosing a house in zone i when having a job in zone j is defined in the usual form as $p_{ij} = p_i \; p_{j/i}$. Now rather than also constraining the model with values of the housing probabilities p_i, we constrain it with knowledge of the entropy S_i of the housing distribution p_i in the form

$$- \Sigma_i \, p_i \, [\log p_i - 1] \; = \; S_i \tag{3.49}$$

In this way, it still remains fundamentally a housing choice model given job, but with the extra knowledge (3.49) of the *uncertainty* of the housing location distribution. With the housing choice entropy given as a constraint, the entropy objective S just contains the average conditional probabilities, yielding

$$S \; = \; - \Sigma_i \, p_i \, \Sigma_j \, p_{j/i} \, [\log p_{j/i} - 1] \tag{3.50}$$

Our knowledge of the job distribution $\overline{p}_j$ is reflected in the following constraints

$$\Sigma_i \, p_i \, p_{j/i} \; = \; \overline{p}_j \tag{3.51}$$

Defining the housing choice problem classically in terms of a trade-off between rents and commuting costs, we add the following constraint on the observed average rents r in terms of the rents r_i in each zone i via

$$\Sigma_i \, p_i \, r_i \; = \; r \tag{3.52}$$

If we have via sampling the average commuting cost c in terms of the interzonal commuting costs c_{ij}, the following constraint is added

$$\Sigma_i \, p_i \, \Sigma_j \, p_{j/i} \, c_{ij} \; = \; c \tag{3.53}$$

The final constraints are the probability normalisation constraints

$$\Sigma_i p_i = 1 \quad ; \quad \Sigma_j p_{j/i} = 1 \tag{3.54}$$

Now, if the entropy objective (3.50) is maximized under the aggregate entropy constraint with multiplier - $(1/\theta)$ [this form is taken to make the nested logit analogy more transparent], the job choice constraints (3.51) with multipliers η_j, the rent constraint (3.52) with multiplier λ, the travel cost constraint (3.53) with multiplier β and the normalisation constraints (3.54) with multipliers Ω and ω_i respectively, then maximization of (3.50) in terms of $p_{j/i}$ yields

$$p_{j/i} = exp - (\eta_j + \beta c_{ij}) / \Sigma_j exp - (\eta_j + \beta c_{ij}) \tag{3.55}$$

If the entropy (3.50) is next maximized in terms of p_i, substituting $p_{j/i}$ from (3.55) and defining 'composite' travel costs $\tilde{c}_i$ as

$$\tilde{c}_i = - (1/\beta) \log [\Sigma_j exp - (\eta_j + \beta c_{ij})] \tag{3.56}$$

the expression for the housing choice probabilities finally emerges as

$$p_i = exp - \theta(\lambda r_i + \beta \tilde{c}_i) / \Sigma_i exp - \theta(\lambda r_i + \beta \tilde{c}_i) \tag{3.57}$$

It is easy to recognise (3.55) and (3.57) as identical in form to a two-level nested logit model (Hensher and Johnson, 1980), where θ is the 'inclusive value' coefficient. If all Lagrange multipliers in both (3.55) and (3.57) are estimated *simultaneously*, the results of our model can be shown to be identical to those of a nested logit model when estimated by the method of *FIML* (Full Information Maximum Likelihood), as pointed out in Roy and Lesse (1985). When θ goes to unity, the model becomes conventional multinomial logit, yielding the joint probabilities p_{ij} upon elimination of η_j as

$$p_{ij} = \tilde{p}_j \; exp - (\lambda r_i + \beta c_{ij}) / \Sigma_i exp - (\lambda r_i + \beta c_{ij}) \tag{3.58}$$

Upon comparison of the simple destination-constrained model (3.58) with that of (3.55) and (3.57), it can be seen that the inclusive value coefficient θ couples the latter set of models, preventing it having the 'naive' IIA structure of (3.58) discussed in relation to the origin-constrained Huff model. Thus, if the number of jobs in any zone j say increases, the relative probabilities of choosing those jobs from the alternative residential zones i will be changed via (3.55) and (3.57), reflecting 'adjacency' influences of those zones i close to the zone j where the jobs are added.

Consider now an attempt to enhance the Huff model by identification of clusters J of retail/service centres, with the choice of an individual centre j being made conditional upon choice of the cluster J where the centre $j \in J$ is located (Roy, 1985). At the same time, we overcome the IIA restriction by introduction of spatial interdependencies via a constraint on cluster entropy, with p_J being the overall probability that cluster J is chosen. Again, we make use of the probabilistic representation, defining conditional probabilities $p_{ij/J}$ as the probability, given that one perceives cluster J, one is from origin zone i and has chosen individual centre j within the cluster. With this definition, the entropy objective S^* is expressed as

$$S^* = - \Sigma_J \; p_J \, \Sigma_{ij} \, p_{ij/J} \, (log \; p_{ij/J} / W_j^{\alpha} - 1) \tag{3.59}$$

With p_i being the exogenous probability that a shopper chosen at random comes from zone i, the origin constraints are given as

$$p_i = \Sigma_{Jj} \, p_J \, p_{ij/J} \tag{3.60}$$

The travel time constraints are given for the average value t in the usual form

$$\Sigma_{ijJ} \, p_J \, p_{ij/J} \, t_{ij} = t \tag{3.61}$$

Then we have the probability normalisation constraints on p_J. The constraint on cluster entropy is given simply as

$$- \Sigma_J \; p_J \, (log \; p_J - 1) = S_J \tag{3.62}$$

Now, if (3.59) is maximized in terms of $p_{ij/J}$, we obtain the following result upon elimination of the Lagrange multiplier on (3.60)

$$p_{ij/J} = p_i \;\; W_j^{\alpha} \, exp - \beta \, t_{ij} / \Sigma_{j \in J} \, W_j^{\alpha} \, exp - \beta \, t_{ij} \tag{3.63}$$

where $j \in J$ represents the set of individual centres j belonging to cluster J. Now returning to the form of $p_{ij/J} = W_j^{\alpha} \, exp - (\lambda_i + \beta \, t_{ij})$ before the Lagrange multiplier on (3.60) was eliminated, we can maximize (3.59) in terms of the cluster probabilities p_j, including constraint (3.61) with multiplier $-(1/\theta)$. If 'inclusive value' cluster travel times t_J are defined as

$$t_J = - (1/\beta) \;\; log \, [\Sigma_{ij} \, exp - (\lambda_i + \beta \, t_{ij})] \tag{3.64}$$

the cluster probabilities emerge in the elegant form

$$p_J = exp - \theta(\lambda_i + \beta \, t_{ij}) / \Sigma_J \, exp - \theta(\lambda_i + \beta \, t_{ij}) \tag{3.65}$$

The above model takes account of 'spatial structure' by partitioning the individual centres into a set of exclusive clusters. By fitting the model to the cluster entropy, the effect of the clustering on choice can be ascertained, that is, how the presence of the cluster influences the choice of an individual centre within the cluster. If a reasonable clustering scheme is not obvious in a particular application, the user can experiment with alternative arrangements. We note that such clustering can pick up the net influences of agglomeration vs substitution. For example, if the centres that are clustered together have quite diverse characteristics, it is likely that agglomeration influences will predominate. As with the previous model (3.55), (3.57), the above cluster model (3.63), (3.65) is also free of the IIA restriction, thus permitting spatial interdependencies to be potentially identified.

3.2.6 Inequality and Logistic Short Run Capacity Constraints

'Capacity' is not always a precise concept in economics. In situations of very high demand, extra capacity can sometimes appear as if 'out of the air'. On the other hand, in short run models, it is usually necessary to apply capacity constraints. In the following, we show that these constraints are defined differently depending on the relative 'homogeneity' or 'heterogeneity' of the regional capacity options that are available from the viewpoint of the occupier of that capacity. We illustrate the concepts by looking at the demand for say IT jobs in an area, where there are more such jobs than workers to fill them. Consider first that the jobs in each zone or region are perceived as almost homogeneous by the workers. Considering a deterministic situation where each worker living in origin zone i chooses a job in any zone j with the maximum unit wage w_j minus generalised commuting costs c_{ij}. This problem would be handled via maximization of the following linear objective

$$W = \Sigma_{ij} T_{ij} (w_j - c_{ij}) \tag{3.66}$$

where T_{ij} are the commuting flows, under the usual origin constraints

$$\Sigma_j T_{ij} = O_i \tag{3.67}$$

where O_i is the number of IT commuters from zone i, and job capacity constraints

$$\Sigma_i T_{ij} \leq S_j \tag{3.68}$$

where S_j is the number of available jobs in zone j, where we are assuming that the total available workers $\Sigma_i O_i$ is strictly less than the total supply of jobs $\Sigma_j S_j$. This linear program will tend to allocate each set of workers to their most preferred job zone given their housing location i [that is, with maximum $(w_j - c_{ij})$], consistent with satisfying the capacity constraints. In practice, other factors will influence job choice decisions and the flow configuration emerging from this solution may be rather different from that observed. In such a case, we turn to an entropy solution, where the objective (3.66) now becomes a constraint on the 'observed' surplus W^0 computed from the observed flows at a chosen base period, expressed via

$$\Sigma_{ij} T_{ij} (w_j - c_{ij}) = W^0 \tag{3.69}$$

As with Wilson (1967), the number of ways (microstates) N that each distinguishable worker within each origin set O_i can choose a job zone j in groups T_{ij} comes out as

$$N== \Sigma_i O_i! / [\Sigma_j T_{ij}!] \tag{3.70}$$

which yields the entropy maximization objective S in the usual way

$$S = - \Sigma_{ij} T_{ij} [\log T_{ij} - 1] \tag{3.71}$$

Now, if (3.71) is maximized in terms of T_{ij} under the origin constraints (3.67) with multipliers λ_i, (3.68) with multipliers η_j and (3.69) with multiplier $-\alpha$, we obtain

$$T_{ij} = O_i \; B_j \exp \alpha(w_j - c_{ij}) / \Sigma_j B_j \exp \alpha(w_j - c_{ij}) \tag{3.72}$$

in which multiplier λ_i has been eliminated by substitution into the origin constraints and $B_j = \exp - \eta_j$. We note that, when any capacity constraint in (3.68) is inactive, that is when some jobs are still vacant, the corresponding Lagrange multiplier η_j equals zero, yielding B_j equal to unity. We also notice, that so long as the capacity constraint in any zone remains inactive, having a zero Lagrange multiplier, it exerts no influence whatsoever on the analysis. At the extreme, if all capacity constraints were inactive, the model would be identical in form to an origin-constrained model (Wilson, 1967).

In a situation where a set S_j of jobs of a given sector is available in a zone, it would generally be expected that, from the standpoint of the workers, some of the individual jobs would be considered as more attractive than others, leading to selectivity on the part of the potential workers of the options *within* each zone (that is, a type of 'Armington assumption'). When this perceived heterogeneity exists, we automatically treat each available job as distinguishable, effectively positing a *distribution* to exist across the available jobs within each zone (Roy and Lesse, 1981). In this case, the number of microstates N^* reflect the number of ways (combinations) the O_i distinguishable workers can be allocated into groups T_{ij} times the number of ways (permutations) the $(\Sigma_i T_{ij})$ workers choosing zone j may select among the S_j distinguishable jobs there, giving

$$N^* = \pi_i [O_i! / \pi_j T_{ij}!] \, . \, \pi_j [S_j! / (S_j - \Sigma_i T_{ij})!] \tag{3.73}$$

where $(S_j - \Sigma_i T_{ij})$ represents the number of vacancies remaining in zone j. Now, if (3.73) is transformed into the entropy S^* in the usual log form using the Stirling approximation, we obtain

$$S^* = - \Sigma_{ij} T_{ij} (\log T_{ij} - 1) - \Sigma_j (S_j - \Sigma_i T_{ij}) [\log (S_j - \Sigma_i T_{ij}) - 1] \tag{3.74}$$

Note that, a sequential dynamic form of this model could potentially trace vacancy chains. As long as the jobs in toto remain in excess supply, we do not require capacity constraints such as (3.68) - the *entropy formulation (3.74) itself* ensures that there are some unfilled jobs in each and every zone!! This is a result of our within-zone job heterogeneity assumption, where there is a 'bottom of the barrel' effect as more and more of the jobs in each zone get taken up. In fact, if one considers the workers arriving incrementally, the zones more favoured à priori, which tend to start to fill first, will exert a *repulsive* effect on future job seekers as they approach capacity and the probability of securing a desirable job falls. If (3.74) is now maximized under the constraints (3.67) and (3.69), we have, after some manipulation

$$T_{ij} = S_j \exp[\alpha'(w_j - c_{ij}) - \lambda_i'] / \{1 + \Sigma_i \exp[\alpha'(w_j - c_{ij}) - \lambda_i']\} \quad (3.75)$$

Expression (3.75) represents a *logistic* job demand function. As pointed out by Johansson (1990), Hotelling (1932) used arguments of regional heterogeneity to justify generic logistic forms of regional supply and demand functions. On the supply side, such functional forms arise in the theory of *vintage analysis* (Johansson, 1990). In contrast to the *separable* demand functions (3.72) resulting from the zonal homogeneity assumption, our logistic function (3.75) is *non-separable*, reflecting the spill-over effects described above. We note (see previous section), that whereas (3.72) has the IIA restriction, (3.75) is non-IIA, with its capability of recognising spatial interdependencies between the level of capacity utilisation in each zone and the relative attractiveness of each competing zone.

3.2.7 Properties of Models under Aggregation

In the early 80's, the author and his colleagues, in Roy *et al.* (1982), used information theory to develop criteria for optimal aggregation of *arrays*. In this section, our purpose is wider. We want to compare calibration and performance of *models* at different levels of aggregation. In cases where we are forced because of problem size or data availability to aggregate up from the level of the individual to a more aggregate level, we want to devise *averaging laws* to enable the results of the more aggregate model to approximate those when the corresponding results of an individual choice model are aggregated up. At the outset, we must confess that this treatment is very partial, and has not been enhanced by empirical research. Thus, it is presented more as an incentive for further work.

Consider a special probabilistic enhancement of the Huff model, where we have a sample of observations on individuals h. At the same time, we would like to add a residential zonal level i, where a subset $h \in i$ of the sampled individuals live. As in the previous section, let us define exogenous probabilities p_i that a shopper chosen at random resides in zone i, conditional probabilities $p_{j/i}$ that given a shopper lives in zone i he is observed to shop at centre j, as well as conditional probabilities $p_{h/ij}$ that given a shopper lives in zone i and shops at centre j, he is individual h within the set $h \in i$. Then, the individual choice entropy S is given as

$$S = -\Sigma_{ij} p_i p_{j/i} \Sigma_h p_{h/ij} [\log p_{h/ij} - 1] - \Sigma_{ij} p_i p_{j/i} [\log (p_{h/ij} / W_j^{\alpha}) - 1] \quad (3.76)$$

where W_j is the retailing floorspace at centre j. The entropy S is maximized simultaneously in terms of $p_{j/i}$ and $p_{h/ij}$ under their normalisation constraints plus a constraint on average travel time t in the form

$$\Sigma_{ij}\, p_i\, p_{j/i}\, \Sigma_{h\in i}\, p_{h/ij}\, t_{hj} \;=\; t \tag{3.77}$$

with Lagrange multiplier β. After some manipulation and re-substitution, the results emerge as

$$p_{h/ij} \;=\; exp - \beta\, t_{hj} \,/\, \Sigma_{h\in i}\, exp - \beta\, t_{hj}\;;\;\; p_{j/i} \;=\; W_j^{\alpha}\, exp - \beta\, t_{ij} \,/\, \Sigma_j\, W_j^{\alpha}\, exp - \beta\, t_{ij} \tag{3.78}$$

in which the 'aggregated' travel times are given in terms of the individual times via

$$t_{ij} \;=\; -\,(1/\beta)\, log\, [\Sigma_{h\in i}\, exp - \beta\, t_{hj}] \tag{3.79}$$

At first sight, (3.79) may seem like the law of averaging which we are seeking. That is, if we have individual travel times t_{hj}, but for reasons of sample size want to calibrate the model at the zonal level i, we may conclude that all we have to do is to compute the averaged travel times t_{ij} and calibrate the model at the usual aggregate level p_i and $p_{j/i}$ of the Huff model. But there are two problems: (i) we do not yet know the impedance parameter β until the aggregate model is calibrated and (ii) the β obtained from calibration of the aggregate model will in general *not coincide* with that which would have been obtained if we had been able to calibrate the full individual choice model (3.78) - it will probably be somewhat larger. However, we can use some initial approximations. For example, with knowledge of values of the gravity exponent n of the Newtonian model with impedance t_{ij}^{-n} (where, for example, n is often about 3 for convenience shopping), we can obtain an averaged value of β in terms of the average shopping time t via

$$\beta \approx n\, log_e\, t \,/\, t \tag{3.80}$$

Thus, we can start by using the above β estimate to compute the appropriate average zonal travel times t_{ij} from (3.79), perform the calibration, obtain a revised value of β, re-compute the travel times, re-calibrate the model and so on. But as we know that this value of β is not the same as if the individual choice model (3.78) itself had been calibrated, we should only carry out an iteration or two.

Thus, whereas (3.79) represents an *elegant log-sum law of averaging*, much more empirical research would need to be carried out to find the value β' which enables the aggregated model to not only approximate the aggregated probabilities of the individual choice model, but at the same time to approximate the average trip time t computed from the base period individual choice probabilities. Thus, the current work here is very modest indeed. At the same time, it represents one of the few efforts to aggregate *models* themselves, as distinct from *arrays* which are part of their input or output.

3.2.8 Surplus Evaluation by Transformation of Entropy Lagrangian

After the identification in the previous sections of the deterministic supply and demand problems of microeconomics as special limiting cases of our probabilistic entropy formulations, it is clear that we can extract most probable (or in most cases 'expected') profits or utility from our widened objective in parallel to average profits or average utility respectively in the classical problem. This then enables us to extract welfare measures from the analysis for project evaluation. The ideas are illustrated by using the supply model formulation in (3.12) to (3.18).

Let us first examine the deterministic profit function (3.13) in terms of revenue determined via the production function (3.12) minus the input costs including transport. Proceeding to our probabilistic formulation, where the entropy function (3.16) is maximised under the row sum constraints (3.15) with Lagrange multipliers λ_{ir} and the 'observed profit constraint (3.17) with multiplier $-\alpha$, the constrained entropy Lagrangian Y, given as

$$Y = -\Sigma_{irs}\, x_i^{rs}\, [\log(x_i^{rs} / F_s) - 1] + \Sigma_{ir}\, \lambda_{ir}\, (X_i^{r0} - \Sigma_s\, x_i^{rs}) - \alpha\, [\Pi^0 - \Sigma_s\, \bar{p}_s\, Z^s + \Sigma_{irs} (p_i^r + c_i^{rs})] \tag{3.81}$$

can be put into the unconstrained form Y' via the following invariant Legendre transforms

$$Y' = Y - \Sigma_{ir}\, X_i^{r0}\, (\partial Y / \partial X_i^{r0}) - \Pi^0\, (\partial Y / \partial \Pi^0) \tag{3.82}$$

where X_i^{r0} represents the observed inputs of sector i from region r and Π^0 the corresponding observed profits. This yields Y' in the form

$$Y' = \alpha\, \Sigma_s\, \bar{p}_s\, Z^s - \Sigma_{irs}\, x_i^{rs}\, [\log\, (x_i^{rs} / F^s) + \lambda_{ir} + \alpha\, (p_i^r + c_i^{rs}) - 1] \tag{3.83}$$

where, as in (3.12) to (3.18), $\bar{p}_s$ is the unit selling price of the output in region s, Z^s is the production function of the output in terms of the input quantities $(\Sigma_r\, x_i^{rs})$ of the intermediate goods i available in region s, F^s is the number of receiving firms producing output in region s, p_i^r are the prices of input goods i from regions r and c_i^{rs} are the transport costs for a unit of good i between regions r and s. Upon examination of the expression (3.13) for the average total profits, it is seen that (3.83) needs to be scaled by the factor $(1/\alpha)$ to yield it as the sum of the total average profits plus a stochastic term. Upon performing this scaling of (3.83) and substituting for x_i^{rs} the entropy maximizing solution (3.18) of our probabilistic problem before the elimination of λ_{ir}, that is,

$$x_i^{rs} = F^s\, \exp\, \alpha\, [\, \bar{p}^s\, \{\partial Z^s(....) / \partial x_i^{rs}\} - \lambda_{ir} - (p_i^r + c_i^{rs})] \tag{3.84}$$

we may then interpret the result as the *most probable maximum profit* (or **producer surplus**) P, expressed as

$$P = (1/\alpha)\ \Sigma_{irs} F^s \exp \alpha [\ \bar{p}_s (\partial Z_s / \partial x_i^{rs}) - \lambda_{ir} - (p_i^r + c_i^{rs})] + \{\Sigma_s \bar{p}_s [Z^s - \Sigma_{ir} x_i^{rs} (\partial Z^s / \partial x_i^{rs})]\} \tag{3.85}$$

in which the final terms in $\{\ ...\}$ brackets would become zero if the production function Z^s were linear, a quite unusual case.

3.2.9 Fulfilment of Key Economic Lemma

One of the fundamental lemmas of microeconomics is Hotelling's Lemma (Binger and Hoffman, 1988). At this point, we will confine the application of this lemma to the supply problem described above - there is an analogous proof on the demand side. The lemma states:

The partial derivative of the maximum profit function with respect to the supply price is identical to the supply function.

With P in (3.85) as the maximum profit function and $\tilde{Z}^s$ as the supply function [that is, the production function Z^s after substitution of the input demand functions (3.84)] and $\bar{p}^s$ as the output price in region s, the Lemma is as follows

$$\partial P / \partial \bar{p}^s = \tilde{Z}^s \tag{3.86}$$

Noting that $\partial x_i^{rs} / \partial \bar{p}^s$ emerges as $\alpha x_i^{rs} \{\partial Z^s / \partial x_i^{rs} + \bar{p}^s (\partial^2 \tilde{Z}^s / \partial x_i^{rs} \partial \bar{p}^s)\}$ and $\partial \tilde{Z}^s / \partial \bar{p}^s = \Sigma_{ir} (\partial Z^s / \partial x_i^{rs}) (\partial x_i^{rs} / \partial \bar{p}^s)$, we may differentiate (3.84) with respect to $\bar{p}^s$ to yield

$$\partial P / \partial \bar{p}^s = \Sigma_{ir} x_i^{rs} \{(\partial Z^s / \partial x_i^{rs} + \bar{p}^s (\partial^2 \tilde{Z}^s / \partial x_i^{rs} \partial \bar{p}^s)\} + \{\tilde{Z}^s - \Sigma_{ir} x_i^{rs} \partial Z^s / \partial x_i^{rs}\} + \bar{p}^s \Sigma_{ir} \{ (\partial Z^s / \partial x_i^{rs}) (\partial x_i^{rs} / \partial \bar{p}^s) - (\partial Z^s / \partial x_i^{rs}) (\partial x_i^{rs} / \partial \bar{p}^s) - x_i^{rs} (\partial^2 \tilde{Z}^s / \partial x_i^{rs} \partial \bar{p}^s)\} \tag{3.87}$$

After considerable cancellation, this yields

$$\partial P / \partial \bar{p}^s = \tilde{Z}^s \qquad \text{Q.E.D.} \tag{3.88}$$

Thus, our probabilistic maximum profit and supply functions, derived from entropy maximisation, satisfy one of the key lemmas of microeconomics. At the same time, we have the advantage of using models fitted to observed average profits, rather than projecting an idealised deterministic profit-maximising solution. A corresponding lemma, but with a negative sign, can be derived analogously for demand functions.

3.2.10 Non-Homogeneous Supply and Demand Functions

In the classical models of microeconomics (Binger and Hoffman, 1988), supply and demand functions are *homogeneous of degree zero*. This implies that if all prices are multiplied by a common ratio (including income in demand models), the quantities will remain unchanged. In turn, this means that we only deal with *relative* prices.

If we examine the functions arising from our probabilistic theory, such as the input demand functions in (3.84), it is clear that the homogeneity property no longer applies. As the prices are arguments of an exponential function, supply or demand quantities will change upon uniform changes in prices. As the prices enter a utility constraint for demand functions and a profit constraint for supply functions, they need in any case to correspond with **market prices**. Otherwise, they would not allow the models to be 'fitted' to *observed* base period average utilities or profits. We do not believe that this difference with the classical models has any fundamental significance - it primarily relates to the appropriateness of the data used to calibrate the models.

3.2.11 Brief Summary

The above framework is the key contribution of this book to the challenge of providing spatial interaction models which are capable of enriching those arising directly from microeconomics. The first step in Sects. 3.2.1 and 3.2.2 is the identification of the microlevel with specific individual agents and their decisions. These are supply agents, demand agents and transformative demand/supply agents making both discrete and continuous spatial choices.

Following the lead set by Smith (1990), the second step in Sect. 3.2.3 becomes the replacement of the objectives of classical deterministic microeconomic theory by equivalent constraints in our probabilistic analogue. We demonstrate the asymptotic properties of the formulation as uncertainty reduces in our probabilistic case and we converge towards the classical deterministic result. The objectives to be maximized embrace profits for the supply side and utility for the demand side.

In Sect. 3.2.4, we examine the information-enriching role of provision of constraints to observations vis à vis the necessity of producing open price-responsive models to enhance the usual spatial interaction models where total supply or demand needs to be given exogenously. This is achieved by adapting a formalism introduced by Lesse (1982). This entails the estimation objective (Lagrangian) being subjected to relevant invariant transformations to produce an 'equivalent' objective which opens up the models for use in projection.

In discrete choice theory, nested model structures arise, such as nested logit. In Sect. 3.2.5, it is shown that similar forms recognising spatial interdependencies can be derived by the introduction of entropy constraints. This approach is related to models developed by Fotheringham in the 80's.

Within a given region, the technology and efficiency levels of the individual firms are rarely identical or even closely so. In Sect. 3.2.6, this fact is accounted for by either defining each unit of regional production capacity as distinguishable from each other, or, equivalently, by defining probability distributions across them. On the supply side, this approach yields the logistic regional supply functions as supported by Hotelling (1932) and given a vintage analysis interpretation by Johansson (1990).

The concluding sections deal with important issues such as the behaviour of the models under aggregation, the fulfilment of Hotelling's Lemma and, last but not least, the corresponding identification of surplus measures for evaluation. The work can be further solidified by generalising existence and uniqueness proofs from those given in Roy and Johansson (1993) for trade flows. The key existence and uniqueness proof, as established by the second author, is given in the Appendix below.

Appendix 3.1

In Roy and Johansson (1993), the following supply function is established for exporters of a given differentiated commodity in region r

$$X_r = \overline{X}_r \exp \upsilon(\pi_r - \tilde{\pi}) / [(1-\rho)/\rho + \exp \upsilon(\pi_r - \tilde{\pi})] \tag{3}$$

where X_r is the volume of exports out of region r, $\overline{X}_r$ is the total export capacity of region r, $\rho/(1-\rho)$ is the total average odds ratio $(\Sigma_r X_r)/[\Sigma_r (\overline{X}_r - X_r)]$ over the system, π_r is the unit profit expressed as $(p_r - \overline{v}_r)$ with p_r the fob price of the good produced by region r and $\overline{v}_r$ the average production costs, υ is an estimated parameter and $\tilde{\pi}$ is an average expected profit given as

$$\tilde{\pi} = (1/\upsilon) \log [(\Sigma_r u_r \exp \upsilon \pi_r)]$$

where u_r is the proportion of the total unutilised capacity of the system which lies in region r. We now quote Proposition 1 without proof.

Proposition 1. *Let the price vector* $p = (p_1, \ldots p_n) \in \{p: p_r \geq 0 \text{ for all } r\}$. *From (3), it is obvious that each of the supply functions can be written as* $X_r(p)$, *i.e. as functions of p. The functions are both continuous in p and take on an ogive* (logistic) *form as* p_r *varies. Moreover, in both cases there is a minimum value* $X_r(p)>0$ *at* $p_r = 0$ *and a maximum value on a selected price space* $0 \leq p \leq \overline{p}$.

In order to examine properties of the exporter/importer equilibrium, the importer demand function is extracted from the paper as

$$x_{rs} = \bar{X}_s \exp \alpha(\pi_{rs} - \Delta\tilde{\pi}_r) / [1 + \Sigma_r \exp \alpha(\pi_{rs} - \Delta\tilde{\pi}_r)] \quad (11)$$

In this expression, x_{rs} is the demand for the commodity from region r in region s, $\bar{X}_s$ is the import capacity of region s, α is a parameter and the link profits π_{rs} are

$$\pi_{rs} = y_{rs} - p_r(1+t_{rs}) - c_{rs}$$

with y_{rs} being the average price paid by the consumers in region s for the differentiated good from region r, t_{rs} is the ad valorem tariff rate for goods from region (country) r entering region s and c_{rs} are the unit transport costs. Finally, the profit correction parameters are given as

$$\Delta\tilde{\pi}_r = (1/\alpha) \log [\Sigma_r (X_s^0/X_r^0) \exp \alpha(\pi_{rs} - \tilde{\pi}_s)]$$

where superscript '0' refers to observed base period values and expected profits $\tilde{\pi}_s$ come out in the form

$$\tilde{\pi}_s = (1/\alpha) \log [\Sigma_r \exp \alpha(\pi_{rs} - \Delta\tilde{\pi}_r)]$$

to be solved simultaneously. We now quote Proposition 2 without proof.

Proposition 2. *The link-specific demand elements $x_{rs}(p)$ in (11) are continuous in p. The aggregate demand $Y_r = \Sigma_s x_{rs}$ directed towards region r has (i) a negative own price derivative, (ii) positive cross price derivatives and (iii) takes on a logistic form as p_r varies. Maximum demand occurs for $p_r = 0$, with all other prices given.*

Now define the excess demand residuals $\bar{Z}_r$ as

$$\bar{Z}_r = (\Sigma_s x_{rs}) - X_r$$

where $(\Sigma_s x_{rs})$ is the amount of the commodity produced at r demanded by all regions s. Let us now quote verbatim the equilibrium condition for this system as given in Proposition 3 of Roy and Johansson (1993).

Proposition 3. *Consider the aggregate supply function $X_r(p)$ in Proposition 1, together with the aggregate demand function $Y_r(p) = \Sigma_s x_{rs}(p)$ in Proposition 2. Assume that $X_r(p) < \Sigma_s x_{rs}(p)$ (i.e. supply is less than demand) at $p_r = 0$ and that $X_r(p) > \Sigma_s x_{rs}(p)$ (i.e. supply exceeds demand) at any price p_r greater than or equal to the maximum $\bar{p}_r$ of the logistic supply curve, as introduced in*

Proposition 1. Then, there exists an equilibrium price p_r, for all other prices given, such that $\bar{Z}_r(p) = \Sigma_s x_{rs}(p) - X_r(p) = 0$. Moreover, there exists a $p^* \geq 0$ such that $\bar{Z}(p^*) = \Sigma_r \bar{Z}_r(p^*) = 0$. The equilibrium is unique.*

Proof. Take all prices except p_r as given. $\bar{Z}_r(p)$ is continuous in p. At $p_r = 0$, $\bar{Z}_r(p) > 0$. $X_r(p)$ approaches the capacity $\bar{X}_r(p)$ as p_r increases. At the same time, $x_{rs}(p)$ approaches zero. Hence, we can select a $\bar{p}_r$ such that $\bar{Z}_r(p) < 0$ at $\bar{p}_r$. Continuity ensures that there must exist a price p_r^* such that $\bar{Z}_r(p) = 0$ at p_r^*. Next, define a $\hat{p}$ vector such that $\hat{p}_1 = \hat{p}_2 = \ldots \geq \bar{p}_r$ for any r. This ensures that $\bar{Z}(\hat{p}) < 0$, while $\bar{Z}(p) > 0$ at $p = 0$. Again we can refer to continuity to prove that there is a p^*, $0 < p^* < \hat{p}$, such that $\bar{Z}(p^*) = 0$.

It is easy to show (see Proposition 2) that (i) each demand function Y_r is strictly decreasing in p_r and (ii) each excess demand function satisfies the gross substitution property (*this is a clear consequence of the strict concavity of our entropy Lagrangian*). Since all supply functions are strictly increasing in their own price and are separable, we can conclude that the price p_r^* such that $\bar{Z}_r(p) = 0$ at p_r^* is unique (Intriligator, 1971). This completes the proof.

As such 'bounded' and continuous supply and demand functions occur throughout this book, the above proof can be regarded as generic.

Part II
Some Examples of Urban Modelling

4 Activities as Generators of Spatial Interaction

4.1 Elements of an Activity Demand Framework

4.1.1 Characterisation of the Demand for Activities

Referring back to Sect. 1.2.3, we see that the paper of Hägerstrand (1970) was pivotal in broadening the concept of the modelling of travel per se towards direct consideration of the spatial consumption of the goods and services (the 'activities') which generate the travel. These activities can be grouped into non-discretionary ones such as work and driving kids to school, and discretionary ones such as shopping and entertainment. This implies, for modelling, that discretionary activities must be allocated over time into the *interstices* separating the non-discretionary activities. It is clear that we have an interlinked process of spatial allocation of both time and money, divided into that needed to travel to and from the activities and that involved in their actual consumption. The understanding and modelling of this set of processes continues to pose problems of combinatorial complexity thirty years after the seminal paper of Hägerstrand. The essential elements of these processes and the resulting implications for modelling are now summarised, drawing extensively on Roy *et al.* (2001).

Changes in Social Activity Behaviour. The distinction between work trips and retail/service trips is being increasingly blurred by recognition of a higher frequency of hybrid multi-stop tours, often including a work stop. A similar 'blurring' occurs with respect to a strict hierarchy of trip types, an essential element of Central Place Theory. Thus, although 'lower order' facilities do not, by definition, possess 'higher order' functions, higher order facilities are increasingly also including lower order functions. Such behaviour is challenging older notions of the separation of activities such as shopping and entertainment with newer notions of "retailment". As peoples' lives are reportedly increasingly time-pressed, they choose activities they can enjoy (Ashley, 1997). Traditionally, most activity modellers, for example, Bowman *et al.* (1998) and Bhat and Koppelman (1993), have categorised activity based on the degree of discretion in undertaking it. A common categorisation is that of Reichman (1976) into Work, Maintenance and Discretionary. The maintenance activities include grocery shopping. Our categorisation defines four distinct tour/order types l on closed circuits, separating pure work-based from pure home-based trip circuits, as well as identifying those work trips which potentially have ancillary retail/service stops. All trips are *potentially* multi-purpose and are defined as follows:

- $l=1$, potential multi-stop tours which start and end at home without a work stop
- $l=2$, potential multi-stop tours which start at home, have a work stop and finally end at home
- $l=3$, potential multi-stop tours which start and end at the place of work, and
- $l=4$, tele-orders

Interdependencies Between Travel and Associated Consumption. Whereas for commuting travel there is a one-to-one correspondence between commuters and jobs, and for commodity flows a one-to-one correspondence between the size of orders and the shipping cost of the commodities, the situation is much more complex for retail/service activity and travel. A typical shopper may potentially make a single trip or multi-stop tour to buy/consume a quite diverse set of commodities/services at different locations in quite variable quantities. At the same time, the general pattern of the tour is clearly dependent on the activities and goods available at potential stops.

Most previous retail/service models estimate the pattern of discrete trips and assign *exogenously* the amount consumed per trip to obtain expenditure at the trip origins and turnover at the trip destinations. This simplification is not too implausible when single-stop home-based or work-based trips are being modelled for a single category of goods or services, such as convenience goods. There is clearly a high probability of repeat patterns on such tours for basic commodities. However, as more multi-worker households arise with coordinated activity patterns, *multi-stop* tours are increasing in frequency. Also, with many large retail centres combining a large variety of goods, services and entertainment under the one roof, any analysis which wants to compare the effectiveness of the current mix of goods/services at a given centre with alternative floorspace policies must clearly be capable of handling *multi-purpose* stops and the potential agglomeration economies thereby available, as stressed by Stahl (1982) and Dellaert *et al.* (1998). The expenditures per trip, treated *exogenously* in existing models, are clearly subject to quite high variation. They are influenced by an increased frequency of multi-stop tours, different retail centre policies on the mix of goods, services and entertainment, changing accessibility standards and travel costs, as well as changing time constraints within households. It is thus clear that a challenge exists to model multi-stop tour patterns (the travel component) *simultaneously* with the multi-purpose money and time expenditure at each facility on the tour where a stop is made (the corresponding consumption component). A proposed 'comprehensive' model makes an initial effort to rise to this daunting challenge. As a first step, we model aggregate discrete tours in parallel with aggregate expenditures on goods and services at each stop, thereby just *loosely* connecting travel and consumption via household budget constraints applied at the trip origins. In the comprehensive model, we attempt to handle the interdependencies between tour and expenditure patterns much more intimately, by simultaneously modelling the tour patterns and the *time and money expenditure **per stop*** on the tours. Being unable to use 'the laws of large numbers' on the expenditures per stop, these must be treated as continuous variables, where integration replaces summation in the entropy objective (Roy, 1997a). This approach is adapted from the continuous space entropy formulation of Angel and Hyman (1972), already discussed in Sec 1.2.4.

Levels of Market Segmentation. One of the current key debates in spatial modelling relates to the use of **aggregate vs disaggregate** models, where the so-called *aggregate* spatial interaction models are often treated dismissively. However, we argue here that this debate is often based on false premises. First, we observe that aggregate models usually divide the analysis into market segments, rather than treating the total population in a single analysis. Note here, that as the number of such segments approaches the number of individuals in our disaggregate sample, the results of the 'aggregate' model become *identical* to those of the corresponding disaggregate model. So long as the market segmentation is cleverly performed such as to minimise the variance of the behaviour within each segment, the results of the two approaches can yield quite close results, even for a modest number of segments. Secondly, some data used in disaggregate models must be extracted from an aggregate analysis. For example, as traffic congestion is an aggregate phenomenon, travel times for individuals in peak periods for a revealed preference analysis must be obtained from an aggregate traffic assignment procedure! Also, because of confidentiality constraints, much of the data on the behaviour of firms is simply unavailable at the level of the individual firm. Finally, although disaggregate demand models have proved their worth for problems with just one or two linked events (eg. joint choice of transport mode and destination), issues of appropriate sample size can become critical for multi-event problems. As shown below, this applies in particular to our models of retail/service activity, where consumers make alternative multi-stop tours jointly with the choice of different quantities of multiple goods and services at each stop. In order to ensure that the number of individuals in each entry of this multi-dimensional choice matrix is statistically not too small, a disaggregate approach must often partition the choice dimensions into several sequential subsets, thereby sacrificing the recognition of interdependencies between the subsets. On the other hand, the more aggregate market-segmented models, such as those presented below, can often retain the full set of choices within the one analysis without the number of entries in each element of the array becoming too small. At the same time, we note that it would be possible, in theory, to analogously define probabilistic *entropy* at the level of the *individual* (Anas, 1983). This would yield an innovative structure similar to logit, but with the addition of our continuous variables, where the 'Gumbel' error terms now become functions of the continuous quantities of good/services/time consumed at each stop. For the reasons stated above, we have deliberately chosen **not** to follow this path.

Market segmentation based on lifecycle is well established in retail modelling, while in the regional science and transportation literature, Hägerstrand (1970) and Jones *et al.* (1983) have noted a very strong influence of lifecycle stage on household activity. Here, the 'market' is segmented into distinct income/life cycle groups, with a separate analysis being performed for each group. The pivotal nature of this influence is also stressed by Wakabayashi and Hewings (2002).

Pointers to Earlier Work. Having defined the main context for the chapter, it is time to point to the large amount of previous work in the fields of retail modelling and activity analysis. Retail and service travel have been represented traditionally by the model of Huff (1964) and its derivatives, as described in Sect. 1.1.2. The Huff model considers home-based single stop travel, being thus incapable of identifying the benefits to shoppers of purchasing or consuming more than one category of goods or services at the one facility. Also, unless augmented by consideration of spatial interdependencies in models such as those of Fotheringham (1983, 1986) and Kitamura (1984), handled in a hierarchical nested clustering formulation by Roy (1985), it cannot even *implicitly* recognise the *potential* for multiple stops on tours. Recent work by Thill and Horowitz (1997) examines the influence of travel time constraints on perceived choice sets. In the first part of this chapter, various extensions of the basic Huff model are discussed, enabling it to be more consistent with microeconomics, as well as giving an improved representation of multi-stop tours. In the second part, a comprehensive activity model is presented, which is still potentially implementable. It attempts to put into practice the activity analysis concepts described already in Sect. 1.2.3 and discussed in more detail above.

4.1.2 The Huff Model and its Extensions

The Usual Form of the Model Plus Some Small Enhancements. Large retail centres, often with catchments strongly overlapping those of their nearest competitors, started being constructed in the late 1950's in large cities in the US and UK. It soon became clear that a model framework needed to be developed to allow projections of the expected viability of the new centres and their influences on existing centres. In response, as already summarised in Sect. 1.1.2, Huff (1964) developed an appropriate model, which has remained remarkably durable. Define O_i as the number of trip origins in residential zone i, e_i the average expenditure per trip from zone i, W_j the total floorspace in facility j, t_{ij} the average travel time between zone i and facility j, n a gravity exponent and S_{ij} the projected flow of revenue between zone i and facility j. Then, the Huff model in its original form projects the flows S_{ij} as the following normalised product in terms of revenue generated, floorspace accessed and a negative power of travel time

$$S_{ij} = e_i O_i W_j t_{ij}^{-n} / [\sum_j W_j t_{ij}^{-n}] \tag{4.1}$$

After the pioneering work of Wilson (1967), a decision-theoretic framework using entropy theory was developed, yielding the number of trips S_{ij} in the form

$$S_{ij} = O_i W_j^{\alpha} exp - \beta t_{ij} / [\Sigma_j W_j^{\alpha} exp - \beta t_{ij}] \tag{4.2}$$

where β is a gravity parameter. The floorspace exponent α represents a scaling effect from the standpoint of customers, estimated simultaneously with β via a linked maximum likelihood criterion. The form (4.2) is the variation of the Huff model most applied in practice. For each major category of goods k, the model (4.2) must be separately estimated.

In order to estimate the gravity parameter β, a survey of shoppers needs to be made to evaluate the average shopping trip time over the entire urban area. The estimation of the customer scale index α requires knowledge of the observed number of trips S_j^0 going to each centre j at the base period. Thus, even this simplest model cannot be estimated on census data alone - it requires data only obtainable from the results of special surveys of shoppers.

The structure of the above form of the Huff model can be readily expanded without making any fundamental changes. For instance, it is possible to make trip generation endogenous and responsive to commonly occurring changes. Trip frequencies can be influenced by transport cost changes, as well as by changes in the number, size and variety of goods in the neighbouring facilities. In order to recognise this, a variation of the usual entropy derivation is made, following Roy and Lesse (1981). If H_i households are resident in zone i, then the entropy of the trip pattern is defined as the number of microstates N generated when the sets of O_i distinguishable trips emanating from each zone i are arbitrarily allocated to the H_i distinguishable households, and then to each centre j, yielding

$$N = [\pi_i H_i^{O_i}] [\pi_i O_i ! / (\pi_j S_{ij} !)] \tag{4.3}$$

where π is the product sign and $!$ the factorial sign. If the log of (4.3) is maximised and the Stirling approximation $x! = x(log\ x - 1)$ applied, together with the usual travel time constraint with Lagrange multiplier β and trip origin constraints $\Sigma_j S_{ij} = O_i$ with multipliers λ_i, the following model emerges

$$S_{ij} = H_i A_i W_j^{\alpha} exp - \beta t_{ij} \tag{4.4}$$

where $A_i = exp - \lambda_i$, taken as invariant from its value $\{O_i^0/[H_i^0 (\Sigma_{j0} W_j^{0\alpha} exp - \beta t_{ij}^0)]\}$ evaluated at the base period, and (W_j^{α}) is introduced as a prior bias factor. In other words, if base period quantities are superscripted by '0', this implies the following expression in terms of base period quantities and projected quantities.

$$(O_i/O_i^0) = (H_i/H_i^0) (\Sigma_j W_j^{\alpha} exp - \beta t_{ij}) / (\Sigma_{j0} W_j^{0\alpha} exp - \beta t_{ij}^0) \tag{4.5}$$

This implies that trip generation is jointly proportional to the number of households and their accessibility to retailing. With this improvement, we are making the first step towards an activity approach, with households introduced as the activity generators. In addition, the first step is also made towards replacing spatial interaction choice models with *endogenous* demand models from microeconomics, one of the main purposes of this book.

If data on the turnover T_j of major categories of goods at major retail facilities j is available, it may be used in place of survey data on the number of shoppers S_j^0 observed at each facility j. Letting e_i be the average expenditure per trip out of each zone i, we may add the following constraints for each facility j

$$\Sigma_i S_{ij} e_i = T_j \tag{4.6}$$

If this constraint with Lagrange multiplier α_j is added to the entropy maximization problem and the floorspace W_j is taken as the prior centre attractiveness, then the model emerges as

$$S_{ij} = O_i W_j \exp(\alpha_j e_i - \beta t_{ij}) / [\Sigma_j W_j \exp(\alpha_j e_i - \beta t_{ij})] \tag{4.7}$$

In this form, the model can project the turnover in future periods, as well as being modifiable (see above) to be household-driven and to endogenise trip frequency.

If we are able to segment the shoppers into income groups l, it makes sense to be more consistent with microeconomics by estimating the model with an expenditure constraint for each such group and major goods category. If e_{il} is the average expenditure on the goods plus travel for members of income group l from zone i, c_{ij} the average travel cost between zone i and facility j and p_{jl} the price of the average 'basket of goods' purchased by a member of income group l at facility j, then the following constraint with multiplier γ_{il} is applied for each zone and income group

$$\Sigma_j S_{ijl} (p_{jl} + c_{ij}) = O_{il} e_{il} \tag{4.8}$$

The addition of constraints (4.8) finally yields the model

$$S_{ijl} = O_{il} W_j^{\alpha} \exp - [\gamma_{il} (p_{jl} + c_{ij}) + \beta_l t_{ij}] / \{\Sigma_j W_j^{\alpha} \exp - [\gamma_{il} (p_{jl} + c_{ij}) + \beta_l t_{ij}]\} \tag{4.9}$$

Some Contributions by Fotheringham and Kitamura. Commencing in the early 80's, Fotheringham (1983, 1986) made some quite far reaching contributions to spatial interaction modelling, as already mentioned in Sect. 1.3.6 and recapitulated here. One of his major theses, as discussed earlier in this book, is that spatial choices are not independent of spatial 'structure', that is, the pattern of spatial opportunities in a given area. If this is true, then conventional gravity models, such as the origin-constrained Huff model used for retailing studies, which have an independence structure, can be partly mis-specified. For instance, when a system with say two identical retail centres spaced well apart is enhanced by the addition of an identical third retail centre, will the *relative* competitive position of the two existing retailers be influenced by alternative location options of the third centre? The conventional model would say no! However, what if the third retail centre were located quite close to say centre number *1*? If each centre were equally accessible to customers, then the conventional model would indicate that addition of the third centre would change the previous market shares from *{1/2, 1/2}* to *{1/3, 1/3, 1/3}*, with no effect on the relative competitive position of the original two centres. Intuitively, for this case of identical centres of equal accessibility, one may expect the market share of centre *1* to be split with that of the new centre, leading to shares of *{1/4, 1/4, 1/2}*, as illustrated in Fig. 1.2. The appreciation of this anomaly gave Fotheringham the incentive to develop new forms of spatial interaction models, which are in some sense hierarchical, as given explicitly in Fotheringham (1986). In other words, customers are not only attracted to a particular centre per se, but to the *milieu* of that centre with respect to adjacent centres. In this way, both spatial competition and spatial agglomeration influences can be identified. It should also be noted, that inspired by Fotheringham's work in 1983, Roy (1985) developed via entropy a hierarchical destination choice or 'cluster' model, which shares the structural dependence properties of the *nested logit* random utility model and overcomes the above IIA anomaly. A nested choice is made of a cluster of centres and a single centre within the cluster.

In illustrating Fotheringham's ideas, we turn to his new competing destinations retail model. In contrast to the conventional retail model in (4.2), the competing destinations model is expressed as

$$S_{ij} = O_i W_j^{\alpha} A_j^{\gamma} exp - \beta t_{ij} / [\Sigma_j W_j^{\alpha} A_j^{\gamma} exp - \beta t_{ij}] \quad (4.10)$$

with the new 'potential' measure A_j defined for 'competing' destinations adjacent to the primary destination j as

$$A_j = \sum_{l \neq j} W_l t_{jl}^{-\beta} d_{jl} \quad (4.11)$$

where t_{jl} represents the unit travel time between the *primary destination* j and a *nearby destination* l. The binary variables d_{jl} are unity when destination l lies within a pre-specified range of destination j and zero otherwise. The introduction of the 'competing destinations' potential terms A_j^{γ} distinguishes the new model from the conventional model in (4.2). A positive value of the scaling index γ demonstrates the presence of consumer agglomeration economies, with a negative value indicating the presence of consumer competition or congestion forces.

The above model structure clearly represents an important step in recognition of interdependencies in spatial choice. It is also supported by a great deal of empirical work by Fotheringham and others. At the same time, there are a couple of 'cosmetic' issues which deserve attention, such as (i) the absence of a floorspace scaling index α on the floorspace W_l of the competing destinations l and (ii) the use of the power form for the impedance of the competing destinations, rather than the exponential form used for the primary destinations j. A remaining challenge is to extend the work of Thill (1995) to formally develop the concept of competing destinations in the framework of multi-stop activity analysis, which is certainly *implied* by the structure of Fotheringham's models. In fact, such an extension would have much in common with the explicit activity analysis framework of Kitamura (1984). As distinct from *trip chain circuits*, the treatment of a primary destination with ancillary secondary destinations, present in the work of both authors, would imply a *star pattern* of trips out from the primary destination, as occurring, for instance, at major subway junctions. This is an alternative view to the continuous trip chain treated here and elsewhere, which is more characteristic of car travel. On the other hand, a more sophisticated approach could regard star patterns of short trips as *intra-tour* trips within a continuous round-trip tour.

A Two-Step Model Recognising Stops on Tours. It is clear that, from an activity context, facilities are accessed via alternative circuits leaving and finally returning to the trip origins i. Thus, it is plausible to initially model tours on such circuits in terms of *aggregate properties* of the paths, such as total path travel time and the total floorspace lying along each path. Then, in a second step, the tours can be allocated to *stops and bypasses* at each facility on the path, in terms of the relative attractiveness of the different facilities there.

Let n_i be a plausible set of such paths accessed from each origin i with circuit times t_{in}. Also, as depicted in Fig. 4.1, let S_{in} be the number of tours on path n_i, V_{ijn} the number of those tours which stop at facility j along the path and W_{in} the total floorspace of all facilities along the complete path.

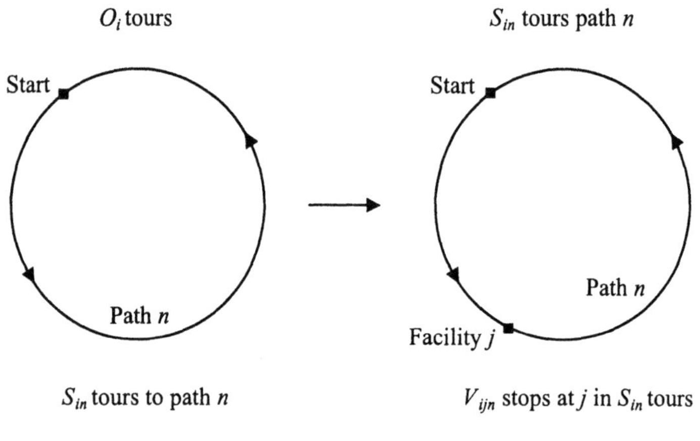

Fig. 4.1 Tours to Paths to Stops

Assuming that W_{in} represents a general attractiveness measure for path n_i, the path tour entropy is written as

$$S_p = - \Sigma_{in} S_{in} [\log(S_{in}/W_{in}{}^{\alpha}) - 1] \tag{4.12}$$

When (4.12) is maximized under the tour origin constraint $\Sigma_n S_{in} = O_i$ and the total travel time constraint $\Sigma_{in} S_{in} t_{in} = O\, t$, where index n refers to the paths n_i, O is the total number of tours and t the average travel time over the system, we obtain

$$S_{in} = O_i W_{in}{}^{\alpha} exp - \beta t_{in} / [\Sigma_n W_{in}{}^{\alpha} exp - \beta t_{in}] \quad (4.13)$$

which is a tour-based version of the Huff model (4.2). Similarly as for the Huff model, trip frequency can be endogenised [see (4.4)-(4.5)].

The next step is to allocate stops and bypasses to all facilities j on each path n_i. Thus, for each such facility j, we allocate the stops V_{ijn} and bypasses $(S_{in} - V_{ijn})$, yielding the following logistic facility entropy

$$S_f = - \Sigma_{ijn} V_{ijn} [log (V_{ijn} / W_j^{\gamma}) - 1] - \Sigma_{in} (S_{in} - V_{ijn}) [log (S_{in} - V_{ijn}) - 1] \quad (4.14)$$

where the floorspace scaling power γ for stops within a tour differs from its counterpart α for choice of the alternative tours from a given origin. We also need information on the number of stops per tour, averaged over the entire area, thus yielding the total number of stops V to be reproduced by the model. If (4.14) is maximized under the constraint $\Sigma_{ijn} V_{ijn} = V$ with multiplier λ, we obtain

$$V_{ijn} = S_{in} W_j^{\gamma} / [W_j^{\gamma} + exp\, \lambda] \quad (4.15)$$

Thus, equations (4.13) and (4.15) are solved in sequence, yielding not only the tour patterns out from each origin, but also the patterns of stops and bypasses at the facilities around each tour circuit. As in (4.8), we should add an expenditure constraint at each zone and perform an analysis *for each household category* l, trading off travel cost and the costs of goods/services consumed, yielding the constraints

$$\Sigma_n S_{iln} c_{in} + \Sigma_{jn} V_{ijln} p_{jl} = O_{il} e_{il} \quad (4.16)$$

where p_{jl} is the price of the average 'basket of goods' purchased at a stop at facility j by a member of a household of class l. Moving even closer to the spirit of activity analysis, we may replace the travel time constraint introduced below (4.12) by a household time budget constraint, defined at each zone i for each household category l in the form

$$\Sigma_n S_{iln} t_{in} + \Sigma_{jn} V_{ijln} t_{jl} = O_{il} b_{il} \quad (4.17)$$

in which, for each household category l, t_{jl} is the average time spent on the shopping/service task at facility j and b_{il} is the combined time budget per household for shopping plus travel in zone i. Note that, all tours consume travel time and cost, whereas only the stops within each tour consume money and visiting time. Also, we note that (4.16) and (4.17) are 'mixed' constraints, containing the tours S_{iln} and the stops V_{ijln} at facilities within the tours, thereby *weakly coupling* the analysis in (4.12) and (4.14) and requiring an iterative solution for the Lagrange multipliers on both (4.16) and (4.17).

The above formulation, if augmented by the suggested inclusion of endogenous trip frequencies, is already an activity model in the true sense. However, we have not yet included *multi-purpose shopping*, preventing identification of the associated agglomeration influences. The models above are just applicable for one type of goods/services. At the same time, we have *endogenised* tour frequencies whilst *exogenising* the expenditure p_{jl} on a given 'basket of goods' (quantity of commodities) consumed per stop for each household category l at each facility j on a tour. However, these variables are clearly interdependent. The comprehensive model below treats multi-purpose shopping, as well as defining the quantities of goods and services consumed per stop as *endogenous continuous* random variables, consistent with the principles of demand theory in microeconomics.

4.1.3 A Comprehensive Activity-Based Model

In this comprehensive model (Roy *et al.*, 2001), the advances included in the recent paper by the first author Roy (1997a) are retained. These include (i) treating trip/order frequency as endogenous for both the retail/service trips and tele-orders, (ii) satisfying for each market segment an average discretionary expenditure constraint containing savings as the numeraire good, (iii) satisfying for each market segment an average time budget constraint having 'other leisure time' as the numeraire good and, last but not least (iv) associating each stop at a facility with the unknown continuous amounts of goods, services and visiting time consumed there. This treatment of multi-purpose stops for consuming different goods/services at any facility potentially identifies between-goods agglomeration influences there. The 'variety' within any one good/service at a facility is identified via maximum likelihood estimation of a power on the floorspace allocated to that good or to any other appropriate 'size' unit for services.

At each facility on a given circuit, for each set of households resident in a given zone and working in the same or another zone, the number of tours around the circuit which by-pass the facility without stopping is distinguished from the number which make a stop. Thereby, the immense combinatorial demands of identifying all possible multi-stop trip chains for each tour on each circuit are avoided. At the same time, continuous quantities of visiting time and goods and services consumed can be associated with each stop, and thus with each multi-purpose, multi-stop tour of each circuit category. This copes with the *'search for variety'*, as also stressed in current trade theory. With such a structure, response of tour frequencies, and thus the total transport task, to both transport policies and facility investments can, in principle, be assessed.

Note that, whereas trip generation for $l=2$ (that is, work trips) is treated as exogenous, these tours are allocated endogenously by the model into pure home-work-home trips and tours which contain one or more extra stops, either on the way to or on the way home from work or both. Trip generation for the other tours $l=1, l=3$ and $l=4$ is regarded as endogenous, with those for $l=1$ and $l=4$ arising from households and those for $l=3$ from workplaces. For any facility j lying on any defined circuit, we must distinguish for each residence/work zonal pair the tours actually stopping at j from those that merely by-pass j on the way to somewhere else on the circuit. At the same time, our model also allows the time spent at a facility to be endogenous, with this time likely to increase as the stop becomes more multi-purpose. Such time, or that spent in making a tele-order, competes via the time budget with all 'other leisure time' not included in the set of activities being modelled. Also, with the inclusion of both expenditure and time budget constraints, the following analysis and model estimation must be carried out in turn for **each** income and/or life style category chosen to stratify the consumers. The results are then easy to aggregate for policy guidance.

Notation. Let there be K_j goods/service activities at any facility j, with any typical activity denoted by k_j, supplemented by one time category for the (endogenous) total time actually spent at the facility during any one stop. The time spent in making a tele-shopping order is taken as a linear function of the total amount of goods ordered in any one call, noting that only a subset $K_j' \in K_j$ of the K_j good/service categories may be available for tele-shopping at a given facility j.

Let S_{ilmn} be the discrete number of *tours* generated by residents of zone i who work in zone m for tour class l on alternative round trip circuits n_{il} or n_{ml}. Adapting the ideas of Angel and Hyman (1972) on continuous space to continuous quantities consumed per trip, let $V_{ijlmn}\ (x_{ij1lmn},\ldots x_{ijklmn},\ldots x_{ijKlmn},\ v_{ijlmn})$ be the *stop/order intensity* at facility j within an (i,l,m,n) tour, as a *function* of the *quantities* x_{ijklmn} (in appropriate units) of activities k consumed during a stop/order at facility j on a tour/order of class l by residents from zone i who work in zone m, and of v_{ijlmn}, the total *visiting time* spent during such a stop at facility j, summed over all the sampled activities k. In order to obtain the total number V_{ijlmn} of *stops/orders* made at any facility j during each such tour, the tours and stop/order intensities must be integrated between the upper and lower limits of goods/services purchased during any one stop/order. When the model is constrained such that it reproduces the total observed turnover (revenue) $R_{jlk}{}^0$ in the base period of each activity k of type l at facility j, we may elect to use the simple integration quantity limits of 0 and ∞. Although this implies that the upper tail of the model distribution extends further than in practice, that is from the maximum expected quantity of each good/service consumed per stop towards infinity, the turnover constraints 'centre' the distribution on the observed average values whilst accepting some approximation in the shape of the distribution. At the same time, the resulting definite integrals then yield an elegant result. The total number of stops/orders V_{ijlmn} is then

$$V_{ijlmn} = \underbrace{\int_0^{\infty} \ldots\ldots\ldots\ldots \int_0^{\infty}}_{K+1\ \text{times}} V_{ijlmn}\,(x_{ij1lmn},\ldots x_{ijklmn},\ldots x_{ijKlmn},\ v_{ijlmn})\ dx_{ij1lmn}\ \ldots\ etc. \quad (4.18)$$

For tele-orders ($l=4$) circuits are irrelevant, and V_{ij4m} is the number of tele-orders made at facility j by individuals resident in zone i and working in zone m. Also, possible constraints on observed actual turnover $R_{jkl}{}^0$ of each good/service type k at each facility j for tours or orders of class l are written as

$$\sum_{imn} \int_0^{\infty} \ldots \ldots \int_0^{\infty} V_{ijlmn} (x_{ij1lmn}, \ldots x_{ijklmn}, \ldots x_{ijKlmn}, v_{ijlmn}) \; p_{jk} \; x_{ijklmn} \; dx_{ij1lmn} \ldots etc. = R_{jkl}{}^0 \tag{4.19}$$

where p_{jk} is the unit price of good/activity k at facility j. If $v_{jl}{}^0$ is the observed average visiting time spent at each facility j (or in telephoning the order when $l=4$) for tour/order category l and $V_{jl}{}^0$ is the corresponding total of stops/orders made at facility j, it is vital to constrain the model to reproduce this observed visiting time via the following integral constraints

$$\sum_{imn} \int_0^{\infty} \ldots \ldots \int_0^{\infty} V_{ijlmn} (x_{ij1lmn}, \ldots x_{ijklmn}, \ldots x_{ijKlmn}, v_{ijlmn}) \; v_{ijlmn} \; dx_{ij1lmn} \ldots etc. = S_{jl}{}^0 \; v_{jl}{}^0 \tag{4.20}$$

As an 'activity analysis' framework must be household-based, let O_i be the exogenous number of households of the current category in zone i. Also, with tour classes $l=2$ and 3 being linked with work locations, let T_{imn} be the (exogenous) number of individual work tours between zones i and m on circuit n_i, subdivided on each circuit n_i for $l=2$ into the set S_{i2mn} which have one or more stops at some facilities on the circuit and those $(T_{im} - S_{i2mn})$ which have no intermediate stops in either direction (that is, pure work trips). With tour generation (except for circuits via work, $l=2$) taken as endogenous, the O_i households are taken to generate, either from home or from work, all the tours/orders associated with zone i. In addition, with s_i taken to be a continuous variable denoting the savings per household in zone i and $E_i(s_i)$ being the intensity of savings as a function of s_i, the following condition must hold, reproducing the number of households O_i generating the savings from the integral over savings intensity

$$\int_0^{\infty} E_i(s_i) \; ds_i = O_i \tag{4.21}$$

As indicated by Morisugi *et al.* (1993), the introduction of such a numeraire good ensures consistency with microeconomic demand theory, allowing the total demand for retail goods and services to be determined endogenously by the model. Note that, the numeraire good can be omitted and the Lagrange multiplier on the base period expenditure constraint treated as a parameter when one is merely examining competition within the retail/service sector. This device ensures that the retail/service expenditure still remains endogenous when using the model for projection. For the time budget constraint a numeraire 'time' good must be introduced, identified here as l_i, a continuous variable denoting the leisure time per household of the current category in zone i, with $F_i(l_i)$ the intensity of leisure at i as a function of l_i, yielding

$$\int_0^{\infty} F_i(l_i) \; dl_i = O_i \tag{4.22}$$

Let W_{jk} be the floorspace or 'size measure' allocated to activity type k in facility j. Also, let $S_{ilmn}{}^0$ be the observed base period tours/orders emanating from people resident in zone i and working in zone m on circuits n_{il} or n_{ml} , $q_i{}^0$ the observed expenditure on all goods/services k per household resident in zone i and v_{ti}^{0} the observed average visiting plus travel time spent per household in zone i. Also, let t_{ilmn} be the observed average return tour time for persons resident in zone i, working in zone m and travelling on circuits n_{il} or n_{ml}, with c_{ilmn} the corresponding return tour cost for tours and c_{ij4} the delivery cost from i to j for tele-shopping. For work trips without a stop, let t_{imn} and c_{imn} be the return travel time and cost respectively between zones i and m on route n_i. Also, let h_i^{0} be the observed expenditure on housing and commuting per household in zone i and e_i^{0} the average amount left for savings from the average discretionary income b^{0}, allocated to goods/services and associated travel, housing/commuting and savings. In addition, f_i^{0} is the observed average leisure time for a zone i household.

Formulation of Entropy Objectives. The entropy microstate approach is used for this model, supplemented by further insights from Roy and Lesse (1981) and Smith (1990), as discussed in Sect. 3.2.1. With the S_{ilmn} tours and the V_{ijlmn} distributions representing the macrostates, microstates are defined to pick up all heterogeneity of individuals and their choice options within the system of macrostates. With all microstates implied by a given macrostate taken as equally probable, the most probable macrostate is that which allows the maximum number of microstates. In applying these ideas to the above problem, we firstly consider the *tour entropy*. Consider firstly the number of ways M_1 that the total number of distinguishable tours S may be allocated into sets S_{ilmn} on each circuit n_i for each (i,m) zonal pair, yielding

$$M_1 = S! / \pi_{ilmn} S_{ilmn}! \tag{4.23}$$

where ! represents the factorial operator and π the product operator. As these tours are generated by households and each household may generate an arbitrary number of tours/orders (except for $l=2$) during the study interval, we express the number of ways M_2 that O_i distinguishable households, subdivided for tour classes $l=2,3$ into their work place locations and routes T_{imn}, can arbitrarily generate distinguishable tours for each class $l=1,3$ via the permutation and combination expressions

$$M_2 = \pi_{imn} \{ O_i^{S_{ilmn}} \} \; \pi_{imn} \{ T_{imn}! / [S_{i2mn}! (T_{imn} - S_{i2mn})!] \} \; \pi_{imn} \{ T_{imn}^{S_{i3mn}} \} \tag{4.24}$$

Similarly, the total savings E can be allocated into zonal groups E_i and then arbitrarily to the households O_i, yielding M_3 microstates as

$$M_3 = \{E! / \pi_i E_i! \} \{ \pi_i O_i^{E_i} \} \tag{4.25}$$

At the same time, an analogous result M_4 occurs for the leisure time F of households

$$M_4 = \{F! / \pi_i F_i! \} \{ \pi_i O_i^{F_i} \} \tag{4.26}$$

Finally, the attractiveness of tours as a whole is represented by allocation of consumers to the heterogeneous selection of goods/services within each category k summed over all facilities passed during the tour. If each unit of available floorspace $W_{ikn} = \Sigma_j W_{jk}$ (where the summation j proceeds over all facilities on circuit n_i)is treated as an elemental choice and each consumer makes just one such elemental choice for each category k, the $\Sigma_{ilmn} S_{ilmn}$ distinguishable tours are allocated to the W_{ikn} distinguishable elemental choices, allowing an arbitrary number of stops for each unit of floorspace, in the following M_5' ways

$$M_5' = \pi_{iklmn} W_{ikn}^{S_{ilmn}} \tag{4.27}$$

In practice, customers may select a partial number of elemental choices within each category k at different frequencies, leading to M_5 total choices in the form

$$M_5 = \pi_{iklmn} W_{ikn}^{(\alpha_{kl} S_{ilmn})} \tag{4.28}$$

where the tour floorspace 'scaling' parameters α_{kl} may be found via a maximum likelihood estimation when the entropy maximization problem is being solved.

The total number of microstates M^T for the tour entropy is expressed as the product $M_1 \bullet M_2 \bullet M_3 \bullet M_4 \bullet M_5$ of the microstates for all the tour processes. Referring to the important *continuous space* paper of Angel and Hyman (1972), M^T is enhanced here by integration over the *continuous quantities* s_i and l_i of savings and other leisure time, yielding consistency with microeconomics whilst maintaining the virtues of our probabilistic entropy approach. Upon taking the natural log of M^T (a monotonic transformation) and again using the Stirling approximation, the discrete/continuous entropy Z^T for tours/orders, facility stops, savings and leisure time is written as

$$\begin{aligned} Z^T &= - \Sigma_{ilmn} \; S_{ilmn} [\log \{ S_{ilmn} / \; [P_{im} (\pi_k \; W_{ikn}^{\alpha_{kl}} \;)]\} - 1] \\ &\quad - \Sigma_{imn} [T_{im} - S_{i2mn}] [\log [T_{im} - S_{i2mn}] - 1] \\ &\quad - \Sigma_i \int_0^{\infty} E_i (s_i) [\log E_i (s_i) / O_i - 1] \, ds_i - \Sigma_i \int_0^{\infty} F_i (l_i) [\log F_i (l_i) / O_i - 1] \, dl_i \end{aligned} \tag{4.29}$$

where P_{im} refers to O_i for $l=1$, unity for $l=2$ and T_{im} for $l=3$. In this expression, **summations** over the **discrete** events are combined with **integrals** over the **continuous** items of the chosen numeraire activities of savings and leisure time, all being evaluated over the survey time interval.

Considering now the *entropy of stops/orders* at facilities, we examine the number of ways M_6 that, for each facility j on the tour, the number of distinguishable tours S_{ilmn} may be allocated to those V_{ijlmn} stopping at j and the number $(S_{ilmn} - V_{ijlmn})$ by-passing j, including tele-orders V_{ij4m} from the total V_4, yielding the logistic entropy (see Sect. 3.2.1)

$$M_6 = \{ \{ \pi_{\substack{ijlmn \\ l=1,3}} [S_{ilmn}! / V_{ijlmn}! \, (S_{ilmn} - V_{ijlmn})!] \} \{ V_4! / \pi_{ijm} V_{ij4m}! \} \} \quad (4.30)$$

where the route index n is irrelevant for tele-orders $l=4$. Using a similar argument to (4.28) to justify scaling factors j on the floorspace W_{ijkn} at each facility j *within* a tour via elemental floorspace choices, the following number of microstates M_7

$$M_7 = \pi_{ijklmn} \, W_{ijkn}^{\{ \gamma_{kl} V_{ijlmn} \}} \quad (4.31)$$

occur, where γ_{kl} is the scaling parameter associated with stops during a tour, which has no direct connection with α_{kl} as used for the tour as a whole.

Realising that the V's are continuous functions of the quantities x consumed per stop/order, the *stop/order* entropy Z^S is finally given as

$$Z^S = - \Sigma_{ijlmn} \int_0^{\infty} \ldots \int_0^{\infty} V_{ijlmn}(x_{ij1lmn}, \ldots, x_{ijklmn}, \ldots x_{ijKlmn}, v_{ijlmn}) [\log \{ V_{ijlmn}(x_{ij1lmn}, \ldots, x_{ijklmn} \ldots x_{ijKlmn}, v_{ijlmn}) / [\pi_k W_{ijkn}^{\gamma_{kl}}] \} - 1] \, dx_{ij1lmn} \ldots dx_{ijklmn} \ldots dx_{ijKlmn} \, dv_{ijlmn}$$

$$- \Sigma_{ijlmn} \; [S_{ilmn} - \int_0^{\infty} \ldots \int_0^{\infty} V_{ijlmn}(x_{ij1lmn}, \ldots, x_{ijklmn}, \ldots x_{ijKlmn}, v_{ijlmn}) \, dx_{ij1lmn} \ldots dx_{ijklmn} \ldots dx_{ijKlmn} \, dv_{ijlmn}] [\log [S_{ilmn} - \int_0^{\infty} \ldots \int_0^{\infty} V_{ijlmn}(x_{ij1lmn}, \ldots x_{ijklmn}, \ldots x_{ijKlmn}, v_{ijlmn}) \, dx_{ij1lmn} \ldots dx_{ijklmn} \ldots dx_{ijKlmn} \, dv_{ijlmn}] - 1]$$

$$- \Sigma_{ij4m} \int_0^{\infty} \ldots \int_0^{\infty} V_{ij4m}(x_{ij14m}, \ldots x_{ijk4m}, \ldots x_{ijK4m}, v_{ij4m}) [\log \{ V_{ij4m}(x_{ij14m}, \ldots x_{ijk4m}, \ldots x_{ijK4m}, v_{ij4m}) \} - 1] \, dx_{ij14m} \ldots dx_{ijk4m} \ldots dx_{ijK4m} \, dv_{ij4m} \quad (4.32)$$

Addition of Information via Constraints to Observations. Commencing first with the tour problem (4.29), we apply the 'definitional' constraints (4.20) and (4.21) on savings and other leisure time. Further constraints are added to enable the model to be estimated according to important observations. Thus, for the model to reproduce the observed number of tours S_{il}^{0} emanating from each zone i, as also required in the classical model of Lakshmanan and Hansen (1966), the following 'origin' constraints are introduced

$$\Sigma_{mn} \, S_{ilmn} = S_{il}^{0} \quad (4.33)$$

Now, we come to the important money and time budget constraints, which ensure that the approach is consistent both with microeconomic theory and activity analysis from the tradition of time geography. As both our 'numeraire' goods (savings and other leisure time) are *aspatial*, separate constraints must be applied at each zone i for both the numeraire and normal goods (Roy, 1997a). Then, in impact analysis, when these constraints must again hold, the two sets are combined into one money budget and time budget constraint applied at each zone. Thus, for each zone, the observed retail/service expenditure q_i^0 per household, including travel costs, must be reproduced by the model via the following constraints

$$\Sigma_{jlmn} \int_0^{\infty} \ldots \int_0^{\infty} V_{ijlmn}(x_{ij1lmn}, .. x_{ijklmn}, .. x_{ijKlmn}, v_{ijlmn}) \, [\Sigma_k \, p_{jk} \, x_{ijklmn}] \, dx_{ij1lmn} .. dx_{ijklmn} ..$$

$$dx_{ijKlmn} \, dv_{ijlmn} + \sum_{\substack{lmn \\ l=1,3}} S_{ilmn} \, c_{ilmn} + \Sigma_{mn} (T_{imn} - S_{i2mn}) \, c_{imn}$$

$$+ \Sigma_{jm} \int_0^{\infty} \ldots \int_0^{\infty} V_{ij4m}(x_{ij14m}, \ldots x_{ijk4m}, \ldots x_{ijK4m}, v_{ij4m}) \, c_{ij4} \, dx_{ij14m} \ldots dx_{ijk4m} \ldots dx_{ijK4m}$$

$$dv_{ij4m} = O_i \, q_i^0 \qquad (4.34)$$

where the delivery cost c_{ij4} for tele-shopping orders is not necessarily an (i,j) cost, but either constant or zero. The observed average zonal savings e_i^0 are reproduced via the integral

$$\int_0^{\infty} E_i(s_i) \, s_i \, ds_i = O_i \, e_i^0 \qquad (4.35)$$

For the time budget constraint, we reproduce the observed total visiting time v_{ti}^0 (including travel) spent per household from each zone i, yielding

$$\Sigma_{jlmn} \int_0^{\infty} \ldots \int_0^{\infty} V_{ijlmn}(x_{ij1lmn}, .. x_{ijklmn}, .. x_{ijKlmn}, v_{ijlmn}) \, v_{ijlmn} \, dx_{ij1lmn} .. dx_{ijklmn} .. dx_{ijKlmn}$$

$$.. dv_{ijlmn} + \Sigma_{lmn} \, S_{ilmn} \, t_{ilmn} + \Sigma_{mn} (T_{imn} - S_{i2mn}) \, t_{imn} = O_i \, v_{ti}^0 \qquad (4.36)$$

Note that, for the tele-shopping orders, when $l=4$, the continuous facility visiting time variable v_{ij4m} is replaced by an *exogenous* observed average tele-order time value v_j or by an *exogenous* proportion of the total order size. Also, the travel time t_{ilmn} vanishes. As for (4.35), the observed average other leisure time per household f_i^0 is reproduced as

$$\int_0^{\infty} F_i(l_i) \, l_i \, dl_i = O_i f_i^0 \qquad (4.37)$$

The observed leisure time left *per household* should be related to just those household members who either tele-shop or who visit facilities - accompanying children would not usually be included except if they were wage earners. At the aggregate level of this analysis, the total household time budget is taken as being consumed arbitrarily by any of the decision-making individuals of that household.

The above constraints are those associated with the *tour* variables S_{ilmn}. However, the money and time budget constraints (4.34) and (4.36) respectively also contain the *stop/order* variables. A further constraint on the *stop/order* problem reproduces the total number of stops/orders V_l^0 observed for each tour class l via

$$\Sigma_{ijmn} \int_0^\infty \ldots \int_0^\infty V_{ijlmn} (x_{ij1lmn}, \ldots x_{ijklmn}, \ldots x_{ijKlmn}, v_{ijlmn})\, dx_{ij1lmn} \ldots dx_{ijklmn} \ldots dx_{ijKlmn}\, dv_{ijlmn}$$

$$= V_l^0 \tag{4.38}$$

If one knows the total number of tours of different categories l, as well as having an estimate of the average number of stops per tour, V_l^0 is readily evaluated. As we will here include powers of floorspace as attractiveness variables, the turnover constraints (4.19) are not applied. However, the visiting time constraints (4.20) are applied as an extra element in facility attractiveness.

Estimation of Entropy Maximization Problems. The *tour* entropy Z^T in (4.29) is maximized in terms of S and the E and F functions under the specified constraints. The Lagrange multipliers associated with the various constraints are to be estimated, and are defined as follows : χ_i with (4.21), τ_i with (4.22), λ_{il} with (4.33), γ_i with (4.34), κ_i with (4.35), μ_i with (4.36) and ξ_i with (4.38). Note that, as no explicit derivatives of E and F functions occur, either in the objective function or the constraints, differentiation under the integral sign may be performed instead of using the more sophisticated variational calculus. The results emerge as

$$\underset{l=1,3}{S_{ilmn}} = P_{im} \{\pi_k W_{ikn}{}^{\alpha_{kl}}\} A_{il} \exp -[\gamma_i c_{ilmn} + \mu_i t_{ilmn}]$$

$$S_{i2mn} = T_{im} \{\pi_k W_{ikn}{}^{\alpha_{kl}}\} A_{il} \exp -[\gamma_i (c_{ilmn} - c_{imn}) + \mu_i (t_{ilmn} - t_{imn})] / \{1 + \exp -[\gamma_i (c_{ilmn} - c_{imn}) + \mu_i (t_{ilmn} - t_{imn})]\}$$

$$E_i(s_i) = O_i B_i \exp - \kappa_i s_i \quad ; \quad F_i(l_i) = O_i D_i \exp - \xi_i l_i \tag{4.39}$$

where $A_{il} = \exp - \lambda_{il}$, $B_i = \exp - \chi_i$ and $D_i = \exp - \tau_i$. In the insertion of the above expressions into the constraint equations to eliminate the Lagrange multipliers, merely simple integrations such as $\int \exp(ax)\, dx$ and $\int x \exp(ax)\, dx$ are required. For instance, the removal of the multipliers on (4.21) and (4.22) yields the following results

$$E_i(s_i) = O_i \kappa_i \exp - \kappa_i s_i \tag{4.40}$$

$$F_i(l_i) = O_i \xi_i \exp - \xi_i l_i \tag{4.41}$$

Except for λ_{il}, the other Lagrange multipliers cannot be eliminated explicitly. Iterative methods such as linear extrapolation or the Newton-Raphson method should be used for this ***estimation*** process, which is complete when all the unknown Lagrange multipliers have been determined.

Now coming to the stop/order entropy Z^S in (4.32), we have a more fully continuous problem. The Lagrange multipliers associated with the various constraints are as follows: η_{jl} with (4.20), γ_i' with (4.34), μ_i' with (4.36) and ϕ_l with (4.38). Upon differentiation under the integral sign, the solution emerges as

$$V_{ijlmn}(x_{ij1lmn}, \ldots x_{ijklmn}, \ldots x_{ijKlmn}, v_{ijlmn}) = S_{ilmn}\{\pi_k W_{ijkn}^{\gamma_{kl}}\} \exp - [(\eta_{jl} + \mu_i')\, v_{ijlmn} + \phi_l$$

$$+ \gamma_i'(\Sigma_k p_{jk} x_{ijklmn})] / \{1 + \{\pi_k W_{ijkn}^{\gamma_{kl}}\} \int_0^\infty \ldots \int_0^\infty \exp - [(\eta_{jl} + \mu_i')\, v_{ijlmn} + \phi_l +$$

$$\gamma_i'(\Sigma_k p_{jk} x_{ijklmn})]\, dx_{ij1lmn} \ldots dx_{ijklmn} \ldots dx_{ijKlmn}\, dv_{ijlmn}\} \qquad (4.42)$$

The integral in the denominator of (4.42), I_{ijlmn}, is readily evaluated, yielding

$$I_{ijlmn} = \exp - \phi_l \,/\, [(\eta_{jl} + \mu_i').\,(\pi_k.\, \gamma_i' p_{jk})] \qquad (4.43)$$

These are some of the key results of the analysis. A similar, but slightly simpler, expression is obtained for the intensity of the tele-orders. By substitution of (4.42) into the relevant constraints, the integration may be carried out analytically before the set of non-linear equations is solved simultaneously for the unknown Lagrange multipliers. Whereas the series of required integrations seems rather tedious, the resulting equations turn out to be relatively simple to solve. Note that, with constraints (4.34) and (4.36) ***coupling*** the tour variables and the stop/order variables, the Lagrange multiplier pairs (γ_i, γ_i') and (μ_i, μ_i') must be adjusted iteratively between the two problems until consistency is obtained.

The final step is to express the total *demand* X_{jk} for goods/services of class k at facility j defined as follows

$$X_{jk} = \Sigma_{ilmn} \underbrace{\int_0^\infty \ldots\ldots\ldots \int_0^\infty}_{K+1 \text{ times}} V_{ijlmn}(x_{ij1lmn}, \ldots x_{ijklmn}, \ldots x_{ijKlmn}, v_{ijlmn})\, x_{ijklmn}\, dx_{ij1lmn} \ldots \text{etc.} \qquad (4.44)$$

in which the stop/order intensity is jointly integrated with the unit quantities x_{ijklmn} demanded per stop. If the intensities are substituted from (4.42) and the integrations performed, the ***demand functions*** emerge as

$$X_{jk} = \Sigma_{ilmn} \frac{S_{ilmn} Z_{ijn} (\exp - \phi_l)\, [\pi_k (1/\gamma_i' p_{jk})]\, [1 + (1//\gamma_i' p_{jk})]}{\{1 + Z_{ijn} (\exp - \phi_l) / [(\eta_{jl} + \mu_i')(\pi_k \gamma_i' p_{jk})]\}\, (\eta_{jl} + \mu_i')} \qquad (4.45)$$

where we have denoted $Z_{ijn} = [\pi_k W_{ijkn}^{\gamma_{kl}}]$. This function, after considerable algebra, can eventually be shown to be strictly decreasing in the prices p_{jk}, as expected from classical demand theory.

The required data include:

- Frequencies of tours and frequencies of stops for three kinds of multi-purpose and multi-stop tours: home-based without involving a work stop; home-based with a work stop; and work-based; given by household;
- Frequencies of tele-shopping orders;
- Tour-based expenditures at facilities, ideally on specific goods/services;
- Tour-based time consumed at facilities;
- Household socio-demographic and socio-economic data;
- Network times and distances for alternative tour paths;
- Locations of facilities and their attributes.

It is clear that special data surveys will need to be performed to collect some of the above data. As some of these items are also required as exogenous input in the myriad applications of the Huff model, the main extra data collection effort relates to identification of the multi-stop tours instead of the earlier single stop tours. Such data is necessary for any plausible implementation of an activity analysis framework.

Setting Up for Impact Analysis. Using the approach of Lesse (1982), it is possible to transform a selected set of the estimated Lagrange multipliers into parameters for using the model in projection, as discussed in Sect. 3.2.1. It is recommended to transform those that represent the *internal state* of the system. In general applications, where the influences of changes in either discretionary income or prices on retail/service expenditure are to be determined, it is necessary to include the 'numeraire' good savings in the analysis, as also stressed by Morisugi *et al.* (1993). In addition, the retail/service expenditure constraint must be merged with the savings constraint into the one constraint on discretionary income, with the savings quantities being scaled by the ratio (κ_i / γ_i) of the respective Lagrange multipliers. However, when the primary objective is to examine competition between retailers, the Lagrange multiplier γ_i, which can be shown as proportional to the price elasticity of demand for households in zone i, can be treated as a parameter in impact analysis. Similar arguments apply for the time budget constraint, although the total time budget for retail/service activity plus other leisure could be rather stable over time. Also, the plausible assumption that the multipliers A_{il} remain reasonably constant enables the tours S_{il} which are generated from each zone, and thus *trip frequency,* to be *endogenous.* As the potential increase of tele-shopping is a dynamic phenomenon not directly identifiable via our comparative static approach, trends are better handled exogenously via a scenario approach, with the model assessing the relative incidence of tele-shopping and the corresponding impacts on profitability in various competing facilities. At the same time, the total level of tele-shopping is responsive to policies such as increases in fuel costs or investments yielding changes in travel time.

As soon as an impact analysis is carried out using the estimated model, many indicators of performance and household welfare can be obtained. These include (i) tour frequency per household from each residential zone, (ii) average tour times on each circuit incurred from each zone and to each retail/service facility, (iii) amount remaining for household savings in each zone, (iv) relative use of tele-shopping vs retail/service tours from each zone, (v) average amount and composition of goods/services purchased per tour from each zone, (vi) revised turnover in each good/service k in each facility due to alternative pricing and supply policies, (vii) price elasticities of demand, (viii) potential profitability indicators in terms of turnover per unit area of floorspace, (ix) facility congestion indicators in terms of customers per unit floorspace and (x) the stop and non-stop patterns of tours at facilities around each circuit. These results are relevant for several policy contexts, including urban transport policy, retail/services pricing policy, decisions on future retail/services capacity including parking, urban environmental policy and policies on household welfare. Their importance is further reinforced by the increasing proportion of work trips involving shopping, changing the focus from pure journey-to-work trips, the major focus of traditional modelling studies.

4.2 Supply, Equilibrium and Imperfect Competition

Having presented a short run demand formulation in the preceding section, we now proceed to develop a short run supply model as a prelude to examining the nature of the resulting short run equilibrium.

4.2.1 Short Run Goods/Services Supply Model with Congestion

A deterministic profit-maximizing retail/services supply model was already presented in Roy *et al.* (1985). In order to allow for aggregation error and lack of perfect information, a probabilistic entropy-based approach is introduced in the following section.

In the short run, let the retailers at any facility j supply quantity X^*_{jk} of goods/services of category k at prices p_{jk}. Let $\overline{p}_{jk}$ be the wholesale price of the goods, which may vary by facility size as well as by its access to warehouses. Also, let a_k be the number of workers required to sell one unit of the good k, taken as a linear function for each good. Let w_{jk} be the wages paid to salespersons handling goods/services category k in facility j. Let all other costs, such as unit rents r_j and facility operating costs o_j be taken as linearly related to the floorspace W_{jk}. Thus, the total profits Π_{jk} accruing to centre j for all goods of category k are given as

$$\Pi_{jk} = X^*_{jk} (p_{jk} - \overline{p}_{jk} - a_k w_{jk}) - W_{jk} (r_j + o_j) \tag{4.46}$$

Facility congestion is introduced by specifying the maximum amount n_k of goods/services per unit floorspace which can be either physically accomodated (goods) or serviced (services). This yields a maximum supply capacity of $\boldsymbol{X}_{jk} = n_k W_{jk}$. At the base period, let the total quantity X^{*0}_{k} of goods of category k be allocated into amounts X^*_{jk} to each facility j. When each observed unit of goods X^{*0}_{k} of category k is taken as distinguishable, this yields M_1 microstates (combinations) of individual allocations X^*_{jk} via

$$M_1 = \pi_k \{ X^{*0}_{k}! / [\pi_j X^*_{jk}!] \} \tag{4.47}$$

where $!$ is the factorial sign. Also, the distinguishable goods/services quantities X^*_{jk} are allocated within the distinguishable total capacity units $\boldsymbol{X}_{jk}$, which, according to Roy and Lesse (1981), yield M_2 microstates (permutations) in the *logistic* form discussed in Sect. 3.2.1 via

$$M_2 = \pi_{jk} \{ \boldsymbol{X}_{jk}! / (\boldsymbol{X}_{jk} - X^*_{jk})! \} \tag{4.48}$$

If the log of the number of microstates $M_1 * M_2$ is taken as the entropy, and if the Stirling approximation $x! = x(\log x - 1)$ is applied, we end up with the following Lagrangian

$$Z = \overset{\max}{\{X^*_{jk}, \lambda_k, \eta_k, \psi_j\}} - \Sigma_{jk} X^*_{jk} [\log X^*_{jk} - 1] - \Sigma_{jk} (\boldsymbol{X}_{jk} - X^*_{jk}) [\log (\boldsymbol{X}_{jk} - X^*_{jk}) - 1] + \Sigma_k \lambda_k (X^{*0}_{k} - \Sigma_j X^*_{jk}) - \Sigma_k \eta_k \{ \Pi^0_k - \Sigma_j W_{jk} (r_j + o_j) - \Sigma_j X^*_{jk} (p_{jk} - \bar{p}_{jk} - a_k w_{jk}) \} - \Sigma_j \psi_j (Y^0_j - \Sigma_k X^*_{jk} p_{jk}) \tag{4.49}$$

where λ_k, η_k and ψ_j are unknown Lagrange multipliers, Y^0_j is the total observed turnover at facility j and Π^0_k is the total observed operating profits for goods of category k, as seen from (4.46). Note that, in terms of the guidelines in Smith (1990), interpreted in Sect. 3.2.1, these profits, used here in the *constraints* with respect to observations, represent the *objectives* of the corresponding deterministic supply models from microeconomics. It can be shown that, in the deterministic limit, as the entropy term vanishes relatively with increasing η_k, the above problem converges to that of the conventional profit-maximizing supply model. If $\exp\text{-}\psi_j\, p_{jk}$ is denoted as B_{jk} and $\exp\text{-}\lambda_k$ as A_k, then the maximization process yields the unique result

$$X^*_{jk} = \boldsymbol{X}_{jk} A_k B_{jk} \exp \eta_k [p_{jk} - q_{jk}] / \{ 1 + A_k B_{jk} \exp \eta_k [p_{jk} - q_{jk}] \} \tag{4.50}$$

where q_{jk} denotes the unit costs ($\bar{p}_{jk} + a_k w_{jk}$). This is seen to be a *logistic supply function*, as recommended by Johansson (1991) in the context of vintage analysis and in earlier work by Hotelling (see Sect. 3.2.1). Also, it is a strictly increasing function of price, as expected. In the short run, if certain facilities are operating at close to their capacity limit, the model will tend to direct most new supply to less congested facilities. This means that the supply model is *non-separable*.

From Sect. 3.2.2, the producer surplus Y_{jk} accruing to firms at j due to sales of goods/services of category k, which can be interpreted as 'expected profits', is directly obtained as

$$Y_{jk} = X_{jk} \{ \log\{ X_{jk} / [1 + A_k B_{jk} \exp \eta_k (p_{jk}-q_{jk})]\} - 1 \} / \eta_k \qquad (4.51)$$

As presented in Sect. 3.2.2, the introduction of our *probabilistic* supply model implies that equalisation of (Y_{jk}/W_{jk}), the *expected* profits per unit floorspace, should be used in place of the *deterministic average* equal profits per unit floorspace criterion used by Harris and Wilson (1978). We recommend this as the criterion for *long run* locational Nash equilibrium for facilities with many independent operators under free entry conditions (see Sect. 4.2.3). When such unit expected profits equalize, there will be no incentive for individual operators to change their location. This substitution of expected for average profits will clearly influence the potential bifurcation behaviour in the dynamic adjustments towards long run equilibrium, compared with that in the original paper.

4.2.2 Short Run Demand/Supply Equilibrium

From the demand functions (4.45) and the supply functions (4.50), it is possible to write the zero excess demand relations as

$$X_{jk} - X^*_{jk} = 0 \qquad \forall_{j,k} \qquad (4.52)$$

Examination of the demand and supply functions (4.45) and (4.50) respectively indicates that only the supply function (4.50) is analytically invertible to give price as an explicit function of quantities. The inverse supply function is readily expressible as

$$p_{jk} = q_{jk} + \log [X^*_{jk} / \{A_k B_{jk} (X_{jk} - X^*_{jk})\}] / \eta_k \qquad (4.53)$$

This analytical invertibility may sometimes allow the efficient cobweb procedure (Intriligator, 1971) to be used to find the equilibrium prices and quantities. Otherwise, we must use iterative techniques such as Newton-Raphson iteration.

Following related work on commodity flows in Roy and Johansson (1993), Propositions 3 and 4, it is seen from Appendix 3.1 that, with strictly decreasing demand functions, strictly increasing supply functions and continuity and boundedness in prices and quantities, a unique equilibrium exists. This result holds by inference for the current Walrasian model system.

4.2.3 Movement Towards Long Run Supply Equilibrium

The short run solution may well yield results indicating that some facilities are performing very poorly, and may go out of business unless major changes are made. In any case, wide variations in profitability among competing facilities will almost certainly lead in the long run to changes in location of some of the tenant businesses. As briefly discussed in Sect. 3.1.1 for a **deterministic** supply criterion , one can look at a dynamic floorspace adjustment process, whereby the **expected** profit per unit floorspace for any goods/services category moves towards the Nash equilibrium condition of profit uniformity via decisions of individual retailers. One can make these adjustments iteratively, with each iteration covering an appropriate time interval. The dynamics can be handled using a logistic adjustment process, as proposed for the deterministic case by Harris and Wilson (1978). At the end of each time interval, short run equilibrium must be re-established according to the instantaneous values of floorspace and prices. Also, it is useful to transform the time intervals, floorspace quantities, prices and unit profits into dimensionless form.

From (4.51), one can represent the condition for this long run locational equilibrium from the standpoint of the suppliers as

$$(Y_{jk} / W_{jk}) = c_k \qquad \forall j \tag{4.54}$$

indicating for each goods/services category k that individual retailers will gradually adjust their locations such that the expected profit c_k per unit of floorspace will equalise over all centres j. This can be compared with the simple Harris/Wilson average profit criterion in eq (3.2). Note also, it is usually assumed that there is a given quantity W_k of floorspace of each type k to be allocated, which constrains the floorspace in each centre such that $\Sigma_k W_{jk} = W_k$. The values of the ratios c_k emerge endogenously during the analysis. An alternative approach is just to allocate enough floorspace such that a minimum viable profit level is reached.

One of the most interesting features of the work of Harris and Wilson (1978) was the identification of bifurcations in the approach towards long run equilibrium. These are even more likely to occur in our models, as (i) we treat the more complex expression of unit expected profits rather than unit average profits as the long run adjustment criterion and (ii) with the absence of prices in the model of Harris and Wilson, floorspace alone adjusts in the long run - our model, with its embedded short run price/quantity supply/demand equilibrium process, also changes prices after each long run iteration. Harris and Wilson identified the main bifurcation being between a more dispersed set of facilities and a more concentrated pattern. This is certainly a case of some practical interest. A further influence on location may be the increasing incidence of teleshopping, where deliveries may either be carried out from warehouses adjacent to the larger retail/service facilities, or completely separate locations may be chosen for the warehouses. The warehouses would need to trade off closeness to freight terminals with closeness to expected customers. In such cases, location-specific freight costs would need to be introduced into the long run profit relations for the suppliers of the ordered goods. Finally, in future work, the influence of the fundamental results for long run location modelling in Fujita and Thisse (2002) should be fully taken into account.

4.2.4 Observations on Imperfect Competition

In the above classical equilibrium approach, the firms are simply assumed to be price-takers. This is unrealistic for many parts of the retail/services sector in most countries. Not only do certain niche markets exist, but there is increasing concentration of ownership leading to an oligopoly situation. If our demand functions (4.45) were analytically invertible to express prices in terms of quantities, we would at least have the chance to develop analytical expressions for the new supply functions, where the inverse demand functions would be substituted for the given prices p_{jk} in the supply objective (4.49). Instead, we must resort to the more tedious procedures illustrated below.

Consider the operator of each retail centre j acting as a short run Cournot oligopolist with respect to the mix of goods $\{X_{jk}, \forall k\}$ of each class k which it is to supply to maximise its total operating profits. This implies that operators don't try to second-guess their competitors. In order to solve this problem, we merely need to remove the j summation from (4.49) to express profit maximization by each centre, as well as the λ_k constraints, and to *notionally* regard the prices p_{jk} as given functions of the quantities X^*_{jk} in this process via the inserted inverse demand functions. This yields the implicit result

$$X^*_{jk} = X_{jk}\ B_{jk}\ C_{jk}\ exp\ \eta_k\ [p_{jk}\text{-}q_{jk}] \ /\ \{\ 1 + B_{jk}\ C_{jk}\ exp\ \eta_k\ [p_{jk}\text{-}q_{jk}]\ \} \tag{4.55}$$

where C_{jk} contains the quantities and the price derivatives, given as $exp\ [\Sigma_l\ (\psi_j + \eta_l)\ X^*_{jl}\ \partial p_{jl}/\partial X^*_{jk}\ ]$. If the demand function (4.45) is transposed and expressed as $F_{jk} = 0$, then implicit differentiation yields the own price/quantity derivatives $\partial p_{jk} / \partial X_{jk}$ as $-\ (\partial F_{jk} / \partial X_{jk}) / (\partial F_{jk} / \partial p_{jk})$, plus similar expressions for the cross-derivatives. In this case, the solution is carried out iteratively, with an assumed starting value of the prices p_{jk}^0 and quantities X_{jk}^0 yielding the price derivatives and thus a first step value of the quantities X_{jk}^1. These new quantity values then enable the prices and price derivatives to be updated to p_{jk}^1 via an iterative solution of (4.45) and so on. Conditions for the existence and uniqueness of the solution need to be established. The main point to be made is that, despite the fact that the demand function is not analytically invertible, plausible numerical approaches can be adopted. However, in oligopoly analysis the main questions relate to (i) the reasonableness of the 'myopic' Cournot assumption and (ii) the possibility of coalitions or collusion between operators. For example, in large cities, large retail centres are often operated by the one agent (eg, Safeway), which is interested in overall profit rather than that in each centre individually. There may be cross-subsidisation of less well-located centres by more successful ones. The realistic treatment of these cases is an analytical challenge. Of course, this applies for all considerations of oligopolistic competition, deterministic or stochastic.

The use of the different models will depend on the purpose of the analysis and data availability. For example, the comprehensive model may be able to evaluate tour rates and associated tour consumption for exogenous input into the simpler models.

5 Knowledge Exchange and Agglomeration

5.1 Fundamentals of the Contact Process

As the economy becomes increasingly knowledge-based in most advanced countries, processes of knowledge exchange become pivotal. These issues were taken up in Kobayashi *et al.* (1991) and related to spatial knowledge production in face-to-face meetings via an elegant formulation by Beckmann (1994), as already described in Sect. 2.3.2. Although some classes of knowledge can be exchanged electronically, more complex knowledge exchange processes still appear to require face-to-face meetings. Such meetings generate travel by at least some of the participants, leading to a requirement for spatial models.

In Fujita and Thisse (2002), as already cited, agglomeration processes were described and analysed in a very general sense. In this chapter, we concentrate on agglomeration influences related to knowledge exchange. In this case, regions with higher than average accessibility of their knowledge workers to other knowledge workers in complementary disciplines have agglomeration benefits.

In Kobayashi *et al.* (1998), a fresh conceptual vision was created, in which a preliminary *random matching* process for pairs preceded the consideration of actually making bilateral face-to-face *contacts.* As the matching is most likely to be made by electronic means, the attraction of a region for matching is related entirely to it's 'size', not its location, with spatial friction being neglected. However, the transformation of a successful matching to a realised meeting does consider travel impedance, with a meeting occurring *if and only if both* members of a matched pair perceive it to be a net advantage. In the following section, an entropy approach for both the matching and the contact processes will be derived, as foreshadowed in the original paper. The contact processes will then be generalised to consider multilateral contacts between individuals and between firms, extending work by Roy and Kobayashi (1996).

5.1.1 Preliminary Matching

Let there be a system of n cities or regions, designated $i = 1,n$. For the matching process being studied, let there be a 'population' of N_i distinguishable individuals available for potential matching in each region i. We consider a 'costless' matching process between pairs of individuals, both within and between regions, which is a good approximation for an initial matching via electronic media. It is also relevant for the case when joint attendance at meetings or seminars, already treated as a 'sunk cost', generates the potential for deeper bilateral contacts. Whereas at the most, only one matching is possible between each pair of individuals in the population, any individual may be matched with several others into matched pairs. The total number of matched pairs P_{ij} which are potentially possible *between* regions i and j is $(N_i N_j)$, with the total number P_{ii} *within* region i $[N_i (N_i - 1) /2]$. In practice, let there be x_{ij} of these potential matchings realised. If $q_{ij} = x_{ij} / P_{ij}$ is defined as the probability that any two individuals in region i and region j are going to find themselves matched, and these probabilities are taken as both identical and independent, then we may consider that the probability $p(x_{ij})$ of the x_{ij} matchings being realised between regions i and j is expressed via the following binomial distribution (Kobayashi *et al.* 1998).

$$p(x_{ij}) = {}_{P_{ij}} C_{x_{ij}} \, q_{ij}^{\,x_{ij}} (1 - q_{ij})^{P_{ij} - x_{ij}} \tag{5.1}$$

where $C_{x_{ij}} = P_{ij} ! / [x_{ij} ! \, (P_{ij} - x_{ij})!]$, the total number of combinations of matched distinguishable individuals between regions i and j. Analogous expressions apply for the distribution of matched individuals within the regions.

In order to illustrate the ideas on a small numerical example, we consider region i with 3 individuals $(N_i = 3)$ a, b and c, and region j with 4 individuals $(N_j = 3)$ d,e,f and g. The $P_{ij} = (N_i N_j) = 12$ potential pairs are clearly:

ad ae af ag bd be bf bg cd ce cf cg

The number of ways (microstates) M_{ij} that the potential matchings P_{ij} may be subdivided into a realised set x_{ij} and a non-realised set $(P_{ij} - x_{ij})$ is clearly

$$M_{ij} = P_{ij} ! / [x_{ij} ! \, (P_{ij} - x_{ij})!] \tag{5.2}$$

Now, if, for example, we take the number of realised matchings x_{ij} as 2, the total possible matchings M_{ij} at the microlevel are given from (5.2) as 66. In detail, the microlevel matchings for any of the two pairs are expressed as:

(ad ae) (ad af) (ad ag) (ad bd) (ad be) (ad bf) (ad bg) (ad cd) (ad ce) (ad cf) (ad cg) ; (ae af) (ae ag) (ae bd) (ae be) (ae bf) (ae bg) (ae cd) (ae ce) (ae cf) (ae cg) ; (af ag) (af bd) (af be) (af bf) (af bg) (af cd) (af ce) (af cf) (af cg) ; (ag bd) (ag be) (ag bf) (ag bg) (ag cd) (ag ce) (ag cf) (ag cg) ; (bd be) (bd bf) (bd bg) (bd cd) (bd ce) (bd cf) (bd cg) ; (be bf) (be bg) (be cd) (be ce) (be cf) (be cg) ; (bf bg) (bf cd) (bf ce) (bf cf) (bf cg) ; (bg cd) (bg ce) (bg cf) (bg cg) ; (cd ce) (cd cf) (cd cg) ; (ce cf) (ce cg) ; (cf cg)

By simple counting, it can be confirmed that the number of microstates ( ) in the table sums to the required total of 66.

With our assumed random matching procedure, it is possible to estimate and solve the entropy maximization problem simply, without the usual requirement for a travel cost or time constraint. Let's assume that we have information on the total number of realised interregional matchings X_r and the number of intraregional matchings X_a, yielding the constraints

$$\Sigma_{i,\, j=i+1,n}\, x_{ij} = X_r \quad ; \quad \Sigma_i\, x_{ii} = X_a \tag{5.3}$$

where x_{ii} represent the intrazonal matchings within zone i. We note that a matching between an individual in region i and an individual in zone j occurs automatically in both directions, with no travel being involved. Hence, we limit the interregional indices to the off-diagonal upper triangle of the *(i j)* matrix, as illustrated in the j indices in the first of (5.3).[1] From (5.2), the interregional entropy S_r, after application of the Stirling approximation, becomes

$$S_r = -\Sigma_{i,j=i+1,n} \{ x_{ij} [\log x_{ij} - 1] + (P_{ij} - x_{ij}) [\log (P_{ij} - x_{ij}) - 1] \} \tag{5.4}$$

If the objective (5.4) is maximized in terms of x_{ij} under the first constraint in (5.3), we finally obtain the following result

$$x_{ij} = X_r\, P_{ij} / [\Sigma_{i,j=i+1,n}\, P_{ij}] \tag{5.5}$$

This is the typical result for a random case, where the interzonal matchings are given in terms of the ratio of the *(ij)* potential interzonal matchings and the total potential matchings in the system. Analogous reasoning leads to the result for the intrazonal matchings, using the second constraint in (5.3), as

$$x_{ii} = X_a\, P_{ii} / [\Sigma_i\, P_{ii}] \tag{5.6}$$

Before concluding this section, we must mention the question of obtaining the matchings calibration data X_r and X_a. As matchings are not 'observed', it is not possible to use *revealed preference* information to obtain the calibration data. Rather, we must carry out a *stated preference* survey on a representative sample of the population to discover the total numbers of unique interzonal and intrazonal matchings which the members of the population have experienced in a given context. This will not necessarily be straightforward - many individuals may have already deleted unpromising matchings at an early stage from their files before being approached to recall these instances in the stated preference survey! Nevertheless, random matching via electronic means or via attendance at large meetings usually precedes the making of any face-to-face contacts, and is thus an important behavioural construct.

---

[1] However, when we come later to consider the face-to-face contacts generated between some of the matched pairs, we need to distinguish between those who visit and incur both travel money and time costs, and those who are visited.

5.1.2 Face-to-Face Contacts Among the Matched Pairs

In Kobayashi *et al.* (1998), a utility-based approach was used to transform the matchings into potential contacts. At the same time, elements of an entropy procedure were sketched. In this section, we not only present a complete entropy solution, but also widen the scope of the problem. As before, we define the meeting intention process between matched pairs in terms of quantities m_{ijk}, $k=1,4$, where $k=1$ implies that both members of the matched pair agree to meeting, $k=2$ implies that the member in region i agrees but not the member in region j, $k=3$ implies that the member in region j agrees but not the member in region i and $k=4$ implies that neither member wants to meet. We extend this analysis by examining the way that the m_{ij1} agreements to meet lead to actual visits v_{ij} where the member of the matched pair in i travels to j, and v_{ji} where the member in j travels to i. In this way, the model can evaluate the effects of accessibility on the willingness of a pair to meet (as in the original model), as well as on the frequency of meetings between such pairs, where repeat visits are allowed for. The travel cost constraint then relates to the actual travel consumption during the study period.

In order to calibrate the model, we first need the distribution x_{ij} of matchings from the model in the previous section. This enables us to write the constraints

$$\Sigma_k m_{ijk} = x_{ij} \tag{5.7}$$

It is also assumed that a stated preference survey yields us the total number m_k of each class of meeting intention k via the constraints

$$\Sigma_{i,j=i,n} m_{ijk} = m_k \tag{5.8}$$

Finally, it is assumed we know the total number of physical visits V, yielding

$$\Sigma_{ij} v_{ij} = V \tag{5.9}$$

where v_{ij} is an element of a full (nxn) square matrix.

Now, the number of microstates M_v in this meeting/visit frequency problem represents the number of combinations in which the x_{ij} distinct matchings may be divided into the four groups m_{ijk} of distinct meeting/no meeting categories k times the number of ways the total observed distinct visits V may be subdivided into distinct zonal groups v_{ij} (including intrazonal visits when $i=j$) times the number of ways that the m_{ij1} distinct meeting intentions may be arbitrarily transformed into distinct visits $(v_{ij}+v_{ji})$, yielding the result

$$M_v = \{\pi_{i,j=i,n}\, x_{ij}\,!\,/(\pi_k\, m_{ijk}\,!\}\ \{V\,!\,/(\pi_{ij}\, v_{ij}\,!)\}\ (\pi_{i,j=i,n}\, m_{ij1}^{\,(v_{ij}+v_{ji})}\ \} \tag{5.10}$$

where for the intrazonal visits we just have (v_{ii}) as the exponent in the final term of (5.10). Upon taking the log of (5.10) and using the Stirling approximation, we obtain the following entropy S_v after dropping the constant terms

$$S_v = - \Sigma_{i,\,j=i,n,k}\ m_{ijk}\ [\log m_{ijk} - 1] - \Sigma_{ij}\ v_{ij}\ [\log v_{ij} - 1] + \Sigma_{i,j=i,n}\ (v_{ij} + v_{ji})\ \log m_{ij1} \tag{5.11}$$

The final step is to give the models some economic content by defining the key behavioural constraint. In this initial formulation, we implicitly assume that there is no difference in the average level of 'knowledge' over the individuals in each region. Thus, once the matching distribution $\{x_{ij}\}$ has been evaluated, we assume that it is accessibility alone which determines the potential for both meeting agreements and visits. Thus, with c_{ij} as the generalised cost of travel between regions i and j (that i, both money and time costs, with the latter being paramount for highly paid knowledge workers), and C as the total such cost in the entire system at the base period, we can write the following constraint

$$\Sigma_{ij}\ v_{ij}\ c_{ij} = C \tag{5.12}$$

where only the i travellers incur travel costs on the visits v_{ij} and the j travellers on the visits v_{ji}.

The problem becomes the maximization of (5.11) jointly in terms of the meeting intentions m_{ijk} and the realised visits v_{ij} under constraints (5.7) with multipliers α_{ij}, constraints (5.8) with multipliers λ_k, constraint (5.9) with multiplier γ and constraint (5.12) with multiplier β. If we define $A_{ij} = exp - \alpha_{ij}$, $B_k = exp - \lambda_k$ and $D = exp - \gamma$, it is easy to show that the results come out as

$$m_{ijk} = A_{ij}\, B_k \quad k = 2,4 \ ; \quad m_{ij1} = A_{ij}\, B_1\, exp\ [(v_{ij} + v_{ji}) / m_{ij1}]$$
$$v_{ij} = D\, m_{ij1}\, exp - \beta\, c_{ij} \tag{5.13}$$

Except for the interesting recurrence relations for determining m_{ij1}, this is a regular entropy maximization problem. We note the strong 'coupling' between the positive meeting intentions m_{ij1} and the realised meetings m_{ij} and m_{ji}, whereby the positive meeting intentions x_{ij1} are seen to be exponentially related to the total number of visits $(v_{ij} + v_{ji})$ per positive meeting intention x_{ij1}. In other words, positive meeting intentions are directly related to the expected unit yield of visits, with the latter itself directly related to accessibility, as shown in the last equation of (5.13). This makes the model interesting and worthy of further development.

In using the model for projection, we would obtain the matching distribution x_{ij} from the analysis in the previous section, but would not expect to have either the total travel consumption C nor the intention category totals m_k. Thus, as illustrated in Sect. 3.2.4, we may form the Lagrangian Z_v by tacking on the constraints with their corresponding Lagrange multipliers onto the entropy objective (5.11). We then perform the following Legendre transforms on Z_v for the meeting intention and travel consumption constraints, yielding the revised Lagrangian Z_v' in the form

$$Z_v' = Z_v - C\, \partial Z_v / \partial C - \Sigma_k\, m_k\, \partial Z_v / \partial m_k \tag{5.14}$$

If this expression is now maximized with β and λ_k now treated as parameters, we obtain again the result (5.13), but where β and D are now given from the initial problem and where the revised travel consumption (C') and the intention category totals (m_k') now emerge as output, as expected.

In Kobayashi *et al.* (1998), considerable attention was devoted to evaluating the expected surplus from the contact process. In Sect. 3.2.8, it is shown how a scaling of the transformed unconstrained Lagrangian yields an analogous measure of surplus. Such a calculation is readily performed on the Lagrangian of the above problem, with its wholly linear constraints. As pointed out in Kobayashi *et al.* (1998), the individuals in a pair whose desire for a meeting is not reciprocated by the other member of the pair experience a loss of surplus.

5.2 An Elementary Modelling Framework for Firms

5.2.1 Elements of a Short Run Probabilistic Knowledge Supply Model

In order to approach the problem of knowledge production and transfer within and between firms, we need to generalise the above analysis. We address firms whose primary function is to transform knowledge inputs into knowledge production, expressed by the revenue accruing to the firms. This is an alternative to the daunting task of creating a true knowledge production function, in which a measure would need to be found for the *quantity* of knowledge. At the same time, as shown in Roy and Kobayashi (1996), we also simplify parts of that analysis. In particular, we bypass the matching process and focus directly on modelling the realised contacts. This enables us to use revealed preference data at the aggregate level, obviating the need for a stated preference survey.

In the earlier work of Kobayashi *et al.* (1991), a general conceptual basis was established for studying knowledge transfer and production, based on the following assumptions (i) the knowledge market is oligopolistic, (ii) knowledge production technology is separable from the normal production technology, (iii) wage levels of knowledge workers are not location-specific, (iv) in the long run, re-location costs are significant or prohibitive and (v) knowledge accessibility is also influenced by the quality of the telecommunications network. However, the extension of the two-region model presented in the paper to a calibrated multiregional model would be very difficult.

In contrast, the approach of Beckmann (1994), presented in Sect. 2.3.2, is much more schematic. This model examines direct contacts between researchers. Its main assumptions include (i) a time budget specific to each individual, divided into time working alone and time spent interacting with other individuals, (ii) a symmetric Cobb-Douglas production function with respect to contacts between individuals and (iii) restriction of the analysis to the short run. By using a further probabilistic argument based on a common aspiration level for all participants, Beckmann produces a gravity-type contact model.

Finally, in Mun and Hutchinson (1995), the intraurban location of office firms is addressed. Some relevant properties of the model include (i) a time budget constraint in terms of time spent in making contact visits and time for 'routine' work, (ii) use of a Cobb-Douglas *revenue* function, (iii) costs defined in the form of uniform wages, variable rents and contact travel and (iv) the calculation of long run equilibrium rents together with the corresponding location pattern of the firms. In this paper, there are some unresolved questions on the method of bid rent determination and the model calibration procedure, which seem to be better handled in an earlier paper (Mun and Yoshikawa, 1993). Despite this, the application of the model to an actual city is a considerable advance.

Desirable Model Properties. In the search for a unified model, selected insights from the above four papers are applied and extended. For instance, as in Beckmann (1994), threshold contact benefits are identified via the dual variables, with different thresholds being identified for visitors versus those being visited, which were not clearly distinguished in earlier papers. Thus, in the modelling of actual contacts, any chosen individual at the microlevel can experience four types of behaviour over a given time interval (i) make no contacts, (ii) visit other individuals, (iii) be visited by other individuals and (iv) both visit and be visited. The frequency of visits is determined endogenously. Note also that not all the meetings which occur at a node j will take place at places of employment - some may occur at neutral venues such a seminar facilities.

Within the array of contacts made by individuals employed by firms, we must allow for the occurrence of meetings of different sizes, ranging from informal meetings of two or three people to more formal seminars. As the 'utility' of such meetings to firms will vary with meeting size, the model defines certain meeting size classes, whose influence is included simultaneously in the revenue functions of the firms.

The inability to quantify knowledge output has a non-trivial influence on the structure of the model. With more 'concrete' goods and services, we can define production functions expressing the output quantity as a function of the required input quantities. These production functions can then be calibrated on available quantity data before being used as part of the profit maximization objective in the determination of the input demand functions. However, no such quantity data exists for knowledge inputs and the resulting output, with revenue functions being introduced instead. As revenue is a key part of the profit objective of the firms, which is maximized in yielding the input demand functions, the parameters of the revenue functions cannot be estimated independently, as for quantifiable goods, but need to be estimated *simultaneously* with the derivation of the input demand functions themselves. This calls for an enhancement of the conventional approach.

As knowledge has a partial public goods character, the suggestion of Romer (1986) is adopted by including in the revenue function of the firms not only the influence of the relative knowledge levels of the individuals whom they employ, but also an agglomeration term reflecting the overall knowledge level of each region as a whole, from which all firms can derive some benefit.

In modelling the behaviour of firms, it is very unusual to be able to obtain individual establishment data, despite such data being implied by the classical theory of the firm. Thus, it is necessary to work with aggregate flows and average revenue and cost levels across each employment node in the study area. The implied unrealistic assumption of dealing with spatial and sectoral aggregation of *identical* firms is tempered by the use of a probabilistic approach. The foundations of such an approach, in which insights from the microlevel are incorporated via entropy into a more macro level analysis, are illustrated for the conventional supply problem in Sect. 3.2.3.

Whereas in an interregional model it is important to include the different levels of telecommunications access in each region (Kobayashi *et al.* 1991), the following analysis for a single urban region assumes that telecommunications standards are uniform across the region. Thus, their influence can be included merely via a 'scaling' constant in front of the revenue function. Without this assumption, the influence of different telecommunications standards would have to be incorporated directly *within* the structure of the revenue functions.

The main part of the following development is devoted to a short run knowledge production model. A short concluding section looks at a long run adjustment process, where the unit profits of the firms tend to equalise in the sense of a Nash equilibrium. Movement towards an oligopolistic equilibrium of the property market is considered together with a competitive equilibrium of the labour market. The model structure is illustrated in Fig. 5.1 below.

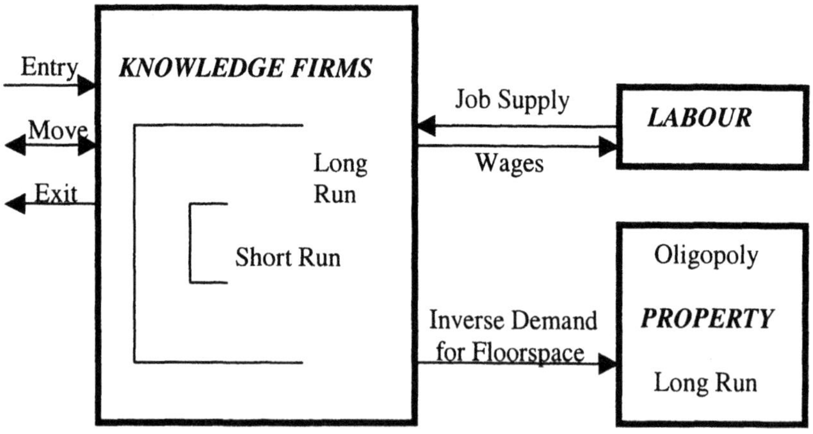

Fig. 5.1 Overall Model Framework

The Statistics of Multilateral Contacts Between Firms. During a representative time period, let contacts between firms be classified by size of meeting into several categories s, where m_s is the average size of a meeting of category s. In such a meeting, m_s contacts are defined to occur, where a contact implies that each individual participant meets the other $(m_s - 1)$ individuals as a group. Let there be N_i representative firms (that is, standardised for size) in each employment node i and let X_{ijs} be the number of individuals from node i visiting node j for meetings of category s. Also, let X_{0js} be the number of individuals *receiving* visits at their own firm in zone j to attend meetings of category s.

As emphasised in Sect. 3.2.1, a key requirement for behavioural use of entropy models is to define the appropriate decision units at the microlevel. In this case, the elemental microlevel decision unit is an individual employed within any of the N_i representative firms at node i who may or may not at this level make contacts or be contacted in any meetings of category s by the $(m_s - 1)$ distinct other individuals in any of the firms N_j at node j, where intranodal meetings are implied when $j=i$. In relation to specifying the problem, it is now necessary to enumerate all these microlevel events which could be generated from any macro pattern of contact flows $\{X_{ijs}, X_{0js}\}$.

Consider a pool of X_s distinct contacts of category s (taken as observed data at the base period) which are to be allocated into groups X_{ijs} from node i to node j. Using standard entropy expressions, this process yields M_1 microstates in the form

$$M_1 = \pi_s X_s! / [\pi_{ij} X_{ijs}!] \qquad (5.15)$$

where $j = 1,n$ for n employment nodes and $i = 0,n$ including the contacts of those being visited. If the representative firms are defined to have k knowledge workers per firm, then $(\Sigma_j X_{ijs})$, $j = 1,n$ distinct *visitor* contacts are allocated arbitrarily to the $(k N_i)$ distinct workers in each origin zone i for each category s, yielding M_2 microstates (permutations) via

$$M_2 = \pi_{ijs} [kN_i]^{X_{ijs}} \qquad (5.16)$$

where (Σ_j) becomes (π_j) when being used as an exponent. Next, at the destinations, $(\Sigma_i X_{ijs})$ distinct contacts (including the visited) of category s are allocated at node j to $M_{js} = (\Sigma_i X_{ijs}) / m_s$ distinct meetings of size m_s (assuming that X_{ijs} is sufficiently large), yielding M_3 microlevel events (permutations) in the special form

$$M_3 = \pi_{js} (\Sigma_i X_{ijs})! / \{ [m_s!]^{M_{js}} \; M_{js}! \} \qquad (5.17)$$

In the above expression, the first term in the denominator represents the M_{js} allocations of individuals into groups of size m_s. However, as we do not want to treat meetings with the same participants as distinguishable, the term $(M_{js}!)$ must be introduced as a further divisor.

Let us illustrate the meaning of (5.17) with an example. Let there be $(\Sigma_i X_{ijs}) = 6$ distinguishable individuals *a, b, c, d, e* and *f* to be allocated to $M_{js} = 2$ meetings of size $m_s = 3$. This yields the following *10* groupings:

abc, def ; abd, cef ; abe, cdf ; abf, cde ; acd, bef ; ace, bdf ; acf, bde ;
ade, bcf ; adf, bce ; aef, bce

From (5.17), the number of groupings is $6\,!\,/\,\{\,(3\,!)^2\,2\,!\,\} = 720\,/\,\{36\,.\,2\} = 10$. Note that, if we had distinguished *[abc, def]*, for example, from *[def, abc]*, we would have had a special case of Boltzmann statistics (see Section 1.2.5), with the same number of individuals (m_s) allocated to each group (meeting). However, as the meeting 'ordering' is inconsequential, the second term in the denominator of (5.17) is required. Note also, that dealing with the large number of meetings taking place in large cities, there is a negligible approximation in the evaluation of the integer number of meetings M_{js} as the quotient $(\Sigma_i X_{ijs}) / m_s$.

Finally, these distinct meetings must be allocated to the host firms N_j at each destination node *j*. However, for the larger meeting size categories *s*, single firms are not typically the hosts. In these cases, a logistic constraint is used (see Sect. 3.2.6) in relation to the total available capacity for such meetings at the node, which is treated as heterogeneous (distinguishable), yielding potential spillover effects into adjacent nodes. In the case of small meetings hosted by the firms themselves, the number of ways M_4 that M_{js} distinct meetings may be allocated to N_j distinct firms is given as

$$M_4 = \pi_{js} \, N_j^{M_{js}} \tag{5.18}$$

Whereas relation (5.16) allows individuals in the origin nodes *i* to make more or less contacts than their fellow workers in the same zones, relation (5.18) potentially allows firms in destination nodes *j* to host more or less meetings than their fellow firms at the same node. This seems a realistic depiction of the microlevel in both cases.

In the system as a whole, the total number of microlevel events *M* is $(M_1 * M_2 * M_3 * M_4)$. The determination of the macrolevel visits $\{X_{ojs}, X_{ijs}\}$ to maximize *M* is our objective. Using again the monotonic log transformation and the Stirling approximation, the entropy expression $S = log\ M$ for the variable terms finally reduces to

$$S = \Sigma_{ijs} X_{ijs} \,[\, log\, \{X_{ijs} /(\, N_i \, N_j^{(1/m_s)} \,)\} - 1\,] \tag{5.19}$$

where, in the parlance of information theory (Snickars and Weibull, 1977) the expression $(N_i \; N_j^{1/m_s})$ in the denominator of the *log* term can be regarded as a *prior probability* of meeting occurrence. Here it is derived endogenously within the microstate evaluation process. Note that, considerable simplification results from constraining the numbers of individuals $(\Sigma_i X_{ijs})$ meeting at node j for meetings of category s to observed values at a chosen base period.

Origin and Destination Flow Constraints. In contrast to deterministic formulations, such as the profit-maximizing choice of inputs by identical firms, the major goal of introducing uncertainty via entropy is an improved empirical performance. As discussed in Sect. 3.2.4, it is vital in model estimation to introduce relevant observed information via constraints which the model must emulate. These can be removed (optionally) to 'open up' the model for projection purposes. If superscript '0' represents observed values, then constraints on the outflows from each origin node i are included via

$$\Sigma_j X_{ijs} = X_{is}^{0} \tag{5.20}$$

where X_{is}^{0} are the observed contacts of category s emerging from node i. Similarly, constraint information is supplied for the destination inflows in the form

$$\Sigma_i X_{ijs} = X_{js}^{0} \tag{5.21}$$

where i is taken to go from 0 to n (including those being visited) and X_{js}^{0} is the total of individuals observed attending meetings of category s in node j.

In the use of the model for projection, new values of X_{is}^{0} and X_{js}^{0} will be clearly unknown. However, by transforming the estimation period Lagrange multipliers on (5.20) and (5.21) to parameters for projection, we can enhance the 'push' factors in (5.19) from merely the number of origin firms N_i and the 'pull' factors from the power N_j^{1/m_s} of the number of destination firms N_j.

Profit Constraints. Whereas the contact costs can be reasonably treated as a linear function of the number of contacts, the revenue functions should be generally regarded as non-linear. As the introduction of non-linear constraints into entropy formulations immediately introduces considerable extra complexity, it costs little to move from the simple Cobb-Douglas revenue functions used by Beckmann (1994) and Mun and Hutchinson (1995) to the somewhat more realistic CES (Constant Elasticity of Substitution) functions.

In Roy (1995), where a probabilistic interregional multi-sector model was formulated, the production functions, representing purely technological quantity-to-quantity transformations, could be assumed to be independently calibrated before being used in revenue constraints in an entropy model based on profit maximization. As the output of knowledge production firms is not *physically* measurable, their parameters cannot be calibrated independently, but must be evaluated *simultaneously* with the Lagrange multipliers of our probabilistic profit maximization model. In this paper, it is recommended that the extra parameters be obtained with respect to a maximum likelihood objective, as already applied to determine the scaling power α in the attractiveness terms W_j^{α} in classical retail models (Roy, 1983). This approach is further reinforced by the intimate connections established between the structure of multilinear models estimated via maximum likelihood and classical entropy models, as demonstrated by Karlqvist and Marksjö (1971).

In defining the revenue function for each knowledge node, the time spent on contacts, the time available for normal work, both subject to a time budget, and the total knowledge level of the region as a whole (Romer, 1986) are each taken to contribute to productivity. The total knowledge level or *stock* of knowledge in the region is taken to be linearly proportional to the average contact frequency per worker, introducing a feed-back effect into the model. The main advances on earlier work include (i) use of the versatile CES function, (ii) the inclusion of different meeting size categories, (iii) the distinction between the visitors and the visited, with the latter not consuming travel time, (iv) the use of a probabilistic rather than a deterministic formulation and (v) the inclusion in the revenue function of the total number of contacts per worker as a proxy for the overall knowledge level for the region as a whole.

For any employment node i, the revenue function R_i can be defined, with t_s being the average time spent per contact of category s, t_{ij} be the average return travel time spent to travel between node i and node j and t_w the working time per individual over the chosen time period of the analysis, yielding

$$R_i = a\, N_i \{ [\Sigma_i\, b_s\, (\Sigma_j\, X_{ijs} + X_{0is})\, t_s / N_i\,]^{-\beta} + [b\, (\Sigma_{ijs}\, X_{ijs}) / N]^{-\beta} + [d\, \{t_w - (\Sigma_{js}\, X_{ijs}\, (t_{ij} + t_s) + \Sigma_s\, X_{0is}\, t_s] / N_i\}]^{-\beta} \}^{-h/\beta} \tag{5.22}$$

where N is the total number of firms $\Sigma_i N_i$ in the region, 'a' is a scaling parameter, b_s, b and d are coefficients, h is the degree of homogeneity and $1/(1 + \beta)$ the elasticity of substitution. In (5.22), the first term expresses the contact time per worker at node i, the second is the average number of contacts per worker over the entire region and the third the time per worker for normal activity at node i.

In order to calibrate the model, a good estimate of the total base period profits Π^0 is required, which the model reproduces in the form

$$\Sigma_i\, R_i - \Sigma_{ijs}\, X_{ijs}\, c_{ij} - \Sigma_i\, N_i\, (k\, w_i\, t_w + u\, n_w\, r_i) = \Pi^0 \tag{5.23}$$

where R_i is the revenue function in (5.22), c_{ij} is the average contact travel cost between nodes i and j, w_i is the average wage rate for knowledge workers at node i, r_i is the weekly rent per unit of floorspace at node i, u is the average floorspace used per firm and n_w is the number of weeks in the observation time period. Thus, in (5.23), the first term is the revenue earned, the second term is the contact travel cost and the third term the total wage plus rent costs.

Solution of Constrained Model. If the entropy (5.19) is maximized with respect to the flows X_{ijs}, with λ_{is} the Lagrange multipliers on (5.20), η_{js} those on (5.21) and α that on the profit constraint (5.23), the solution emerges in the form

$$X_{ijs} = N_i\, A_{is}\, N_j^{1/m_s}\, B_{js}\, exp\, \alpha(\partial \Pi^0 / \partial X_{ijs}) \tag{5.24}$$

where $A_{is} = exp - \lambda_{is}$ and $B_{js} = exp - \eta_{js}$. From (5.22), it is seen that whilst the derivative $\partial\Pi^0/\partial X_{ijs}$ can be evaluated analytically, it remains a function of the flows X_{ijs}. Thus, the solution method needs to be performed via successive substitution. As the CES function in (5.22) is strictly concave, together with the entropy objective itself, the solution remains unique. The 'balancing factors' A_{is} and B_{js} are evaluated by the usual recursive procedure. As the profit Lagrange multiplier α must be determined simultaneously with the revenue function parameters a, b_s, b, d, h and β, the resulting non-linear equations also need to be solved via a well-established iterative method, such as Newton-Raphson.

The maximum likelihood equations for the revenue function parameters are obtained by maximizing the multinomial log-likelihood function F, given as

$$F = \Sigma_{ijs} X_{ijs}{}^0 \log (X_{ijs} / \Sigma_{ijs} X_{ijs}) \tag{5.25}$$

with respect to the revenue function parameters. As (5.24) expresses X_{ijs} as a function of itself, implicit differentiation is required on (5.25) in evaluating such terms as $\partial \log X_{ijs} / \beta$ etc. For instance, after taking the log of both sides of (5.24), subtracting the right-hand side from the left-hand side and denoting the result as G_{ijs}, the derivative with respect to β is written as

$$\partial \log X_{ijs} / \partial \beta = - \partial G_{ijs} / \partial \beta / [X_{ijs} (\partial G_{ijs} / \partial \beta)] \tag{5.26}$$

As the extremal equations will all still contain explicit X_{ijs} terms, the value of X_{ijs} in the previous iteration will be substituted for such X_{ijs} terms within the current iteration.

The arguments from Evans (1973), already discussed in Sect. 3.2.3 can be readily extended from the linear transport cost constraint to our concave profit constraint. As the Lagrange multiplier α approaches infinity, implying a vanishing role for the entropy term in the combined Lagrangian, the solution approaches asymptotically the classical deterministic economic result for profit maximization subject to a CES revenue function and a linear cost function. This is an essential property of our formulation.

In the use of the model for projection, the Lagrange multipliers A_{is}, B_{js} and α in (5.24) are transformed into parameters via Legendre transforms of the entropy Lagrangian (Lesse, 1982), as already adapted to this class of problem by the author in Section 3.2.4. In fact, if no further constraints, such as on labour supply, are to be applied at the projection time period, relation (5.24) with the calibrated values of A_{is}, B_{js} and α can be used directly. As (5.24) is separable except for the CES function influence under the exponential, our model will have similar non-separability properties to its deterministic CES counterpart.

Agglomeration Influences. As in Mun and Hutchinson (1995), it is possible to evaluate the full set of knowledge accessibility *agglomeration* effects by first computing the derivatives $\partial R_i / \partial N_l$, defining the rate of increase of revenue at node i with respect to the number of firms in node l $(l \neq i)$. The implicit algebra required is somewhat simplified by recognition of the fact that $\partial R_i / \partial X_{ijs} = 0$ for $l \neq i$, implying that $\partial R_i / \partial N_i$ is expressed as

$$\partial R_i / \partial N_l = \Sigma_s (\partial R_i / \partial X_{ils})(\partial X_{ils} / \partial N_l) \qquad (5.27)$$

where the latter derivatives are obtained using implicit algebra. Examination of the form of the above derivatives indicates the the extent of the agglomeration economy depends not only on the obvious influence in (5.22) of the total level of contacts in the entire region, but crucially on the degree to which the rate of gain of revenue from the contacts exceeds the rate of loss due to the resulting time subtracted from normal work.

5.2.2 Long Run Adjustment Processes

Location of Knowledge Firms. In the long run, given rents and wages, the firms are assumed to re-locate, or to enter or leave the market to improve their relative position. A long run Nash equilibrium will be posited to occur when a *uniform viable* average profit level per firm is reached across the entire region, whereby if any firm unilaterally tries to change its location, it will be worse off. There are analogies here with the simpler framework of Harris and Wilson (1978). Note also, as discussed in Sect. 3.2.10, with the supply function (5.24) not being homogeneous of degree zero, absolute rather than relative profits are required. Thus, rather than using the classical long run equilibrium criterion of *zero* unit profits at each node, we can set a *uniform viable* level of profits which is typical for the knowledge sector. This fact enables determination of not only the re-location of existing firms, but also the possibility of firms entering or leaving the market such that this viable level of profits is maintained. For simplicity, free entry conditions are assumed for the firms entering the market. This is because the model just determines the new equilibrium location pattern N_i^*, representing the *net* effect of re-location, entry and exit. Thus, it has no way of distinguishing between the actual magnitudes of these individual +ve and -ve effects. In order to obtain such a decomposition, we would need to model dynamically the *gross* movement process.

Let π^* be the relevant viable profit per firm which is to be achieved at each node i via locational adjustments and entries/exits, yielding the relations

$$(R_i - \Sigma_{js} X_{ijs} c_{ij}) / N_i - k\, t_w\, w_i - u\, n_w\, r_i = \pi^* \qquad \forall i \qquad (5.28)$$

from the profit function (5.23). For a typical iteration τ, the non-linear system (5.22), (5.24) and (5.28) can be solved via the following quasi balancing factor approach

$$N_i^\tau = (R_i - \Sigma_{js} X_{ijs} c_{ij}) / (\pi^* + k\, t_w\, w_i + u\, n_w\, r_i) \qquad (5.29)$$

where the right hand side is evaluated at the previous iteration value $N_i^{\tau-1}$. Note that, within this iterative loop, X_{ijs} and R_i must themselves be iteratively adjusted as N_i changes. The non-linearity of both X_{ijs} and R_i in N_i, introduced via the CES revenue function, means that multiple equilibria will exist, as already identified for retailing systems by Harris and Wilson (1978). If a logistic dynamic process of adjustment towards long run equilibrium, as used by them, is adopted, it may be possible to identify significant bifurcations. In the first instance, it would be desirable to simulate this adjustment process for a simple 2-node model, varying the CES function parameters β and h, as well as the unit profit level π^*. We would be seeking situations where small changes in these parameters would yield large changes in behaviour into a quite different equilibrium state.

As suggested by Kobayashi in Roy and Kobayashi (1996), we should, in our probabilistic model, equilibrate on the *expected* unit profit level, rather than the *average* unit profit level used in the deterministic supply function and associated equilibrium solution of Harris and Wilson (1978). As shown in Sect. 3.2.8, this 'producer surplus' level can be evaluated by direct substitution of the solution (5.24) into the entropy Lagrangian, with division by the profit parameter α converting the dimensionless entropy to a surplus value. This transforms the Lagrangian into one term denoting the deterministic profit plus an additional term representing the entropy information. The additional term tends towards zero when its divisor α tends towards infinity, when our probabilistic solution approaches asymptotically the classical deterministic result (Evans, 1973). This short cut approach is possible because of the strict concavity properties of the constrained objective and the resulting uniqueness of the path of the surplus integral, as demonstrated by Champernowne *et al.* (1976). The occurrence of the implicit relationships in both X_{ijs} and R_i means that the surplus does not emerge in the traditional explicit log-sum form, but is itself an implicit function. Thus, the solution algorithm is somewhat more complicated than the relation (5.28), which is based on the equalisation of *average* profit levels. Finally, the implications of the recent results of Fujita and Thisse (2002) on the above locational equilibrium analysis need to be examined in future work.

Nested Labour Market Equilibrium. At the same time that the locations and entries/exits of the firms are adjusting towards a long run Nash equilibrium, wages adjust continuously to match the evolving job demands of the workers, where commuting costs are here accounted for. Let T_{ki} be the number of knowledge workers living in residential zone k and commuting to job node i as part of the $T = \sum_{ik} T_{ki}$ total knowledge workers in the system. Also, let O_k^0 be the base period number of knowledge commuting from zone k and $D_i^0 = k\, N_i$ be the number of base period jobs at node i. If t^0 is the observed average commuting time and s^0 the observed salary net of transportation costs, a slight enhancement of the doubly-constrained model of Wilson (1970) yields the following objective

$$S^c = - \sum_{ik} T_{ki} \,[\log T_{ki} - 1] \tag{5.30}$$

To this objective we attach the classical origin constraints with multipliers λ_k

$$\Sigma_i T_{ki} = O_k^0 \tag{5.31}$$

as well as the classical destination constraints with multipliers η_i.

$$\Sigma_k T_{ki} = D_i^0 \tag{5.32}$$

We also include the classical travel time constraint with multiplier β

$$\Sigma_{ik} T_{ki} t_{ki} = T t^0 \tag{5.33}$$

where t_{ki} is the travel time between zone k and job node i. On the positive side, we add additional income constraints net of travel costs in the form

$$\Sigma_{ik} T_{ki} (w_i - c_{ki}) = T s^0 \tag{5.34}$$

with multiplier $-\nu$, where c_{ki} are the relevant unit travel costs. If (5.30) is maximized in the usual way, the Lagrange multipliers λ_k, η_i, β and ν are evaluated. In the projection period, the new residential worker distribution O_k is provided as exogenous input. On the other hand, the new job distribution and the resulting average travel time and wages net of transport costs should be available as exogenous outputs. This implies that the Lagrange multipliers η_i, β and ν are treated as parameters via Legèndre transforms of the entropy Lagrangian (see Section 3.2.4), yielding the following projection model

$$T_{ki} = O_k B_i \exp [\nu(w_i - c_{ki}) - \beta t_{ki}] / [\Sigma_j B_j \exp [\nu(w_j - c_{kj}) - \beta t_{kj}] \tag{5.35}$$

where the destination attractiveness term $B_i = \exp - \eta_i$. If $(\Sigma_k T_{ki})$ is evaluated from (5.35) as the future number of workers choosing to work at job node i and $(k N_i)$ as the jobs supplied at node i, the wage levels at equilibrium w_i^* adjust to resolve the following excess demand conditions

$$\Sigma_k T_{ki} - k N_i = 0 \qquad \forall i \tag{5.36}$$

at each job node i. If we define g_{ki} as $\exp - (\nu c_{ki} + \beta t_{ki})$, the following simple iterative relationship emerges from (5.36) for the equilibrium wages w_i^*

$$w_i^* = (1/\nu) \log \{ k N_i / [B_i \Sigma_k O_k g_{ki} / \Sigma_j B_j g_{kj} \exp \nu w_j] \} \tag{5.37}$$

As the wages occur on both sides of (5.37), the solution is performed by successive substitution. This short run equilibration procedure must be repeated for each long run change of the location pattern of firms N_i during the iterative adjustment process (5.29).

Note that, whereas accessibility to workers does not enter *directly* into the profit function of the firms, accessibility to jobs enters the worker spatial demand functions (5.35), with potentially higher wages being required in locations less accessible to knowledge workers. Thus, knowledge firms will *in the long run* have an incentive to locate closer to their potential work force, as well as locating *directly* with respect to contacts with other firms. Note also, that traffic congestion has not been endogenously included in the above analysis. A traffic assignment algorithm could be readily included within both the commuting model and the model for work-based contacts. This would influence not only the job choice of workers (and in the long run their choice of housing), but also the contact patterns between the knowledge workers in the firms and their corresponding long run location decisions.

Oligopolistic Office Floorspace Equilibrium. As the commercial property market has a strong oligopoly character in many large cities, it is desirable to go beyond the usual perfectly competitive paradigm. In this exploratory model, we introduce the simplest Cournot oligopoly framework (Intriligator, 1971). As seen in Mun and Hutchinson (1995), we could formulate a short run model in terms of maximization of developer profits for the floorspace actually rented. In this work, we substitute the *long run* inverse demand function for rents in terms of the floorspace demand of the firms. For this purpose, we transpose (5.28) into a *long run* profit criterion for the supply of floorspace by the oligopolistic developer, assumed to be active at each node i. This procedure maintains consistency with respect to speeds of adjustment between the two processes, and is analogous to the retail oligopoly formulation in Roy *et al.* (1985). From Tabuchi (1986) and Mun and Yoshikawa (1993), it is possible to define a long run developer profit function for the provision of office floorspace. If S_i is the office floorspace supplied at node i, A_i is the total developed land for offices at i and l_i the unit land price payed by the developer at i, the total developer profit Π^* is written as

$$\Sigma_i (r_i - o_i) S_i - c A_i (S_i / A_i)^{\beta} - l_i A_i = \Pi^* \tag{5.38}$$

where c and β (>1) are unknown parameters and o_i is the linear-in-floorspace component of operating costs. If u is the average floorspace used per firm, the function $r_i(N_i)$ as extracted from (5.28) is written as $r_i(S_i/u)$ and inserted for r_i in (5.38) for the oligopoly analysis.

We now embed (5.38) into an entropy maximization framework, where distinguishable units of the total base period floorspace S^0 are allocated into nodal clusters S_i. At the same time, with parameters β and c occurring directly in the profit expression (5.38), they must be estimated via maximum likelihood *simultaneously* with the Lagrange multipliers of the entropy formulation, as illustrated already for the revenue function parameters of the firms. The following entropy objective is used

$$S^f = - \Sigma_i S_i [\log S_i - 1] \tag{5.39}$$

The total base period floorspace S^0 must be allocated to nodes i via

$$\Sigma_i S_i = S^0 \tag{5.40}$$

with Lagrange multiplier Ω. Also, (5.38) is applied as a constraint with multiplier ϕ, in which r_i enters as the function $r_i(S_i/u)$. If we denote $[r_i(S_i/u) + S_i\, r_i'(S_i/u)]$, as D_i, where the prime represents the derivative with respect to S_i, the entropy maximization process yields

$$S_i = E \exp \phi\, [D_i - o_i - c\beta (S_i / A_i)^{\beta-1}] \tag{5.41}$$

where $E = \exp - \Omega$. This is again solvable via successive substitution. Note that, D_i is an awkward implicit function of N_i. Simultaneously, via a maximum likelihood objective in terms of S_i, similar to that in (5.25) and (5.26) for X_{ijs}, parameters β and c can be estimated, again using implicit algebra. The maximum likelihood equations come out in the following form

$$\Sigma_i S_i F_i(S_i) = \Sigma_i S_i^0 F_i(S_i) \tag{5.42}$$

where S_i^0 are the floorspace supplies at the base period. It can be checked that for β, the functions $F_i(S_i)$ come out as

$$F_i(S_i) = (S_i/A_i)^{\beta-1} [1 + \beta \log (S_i/A_i)] / [1 + \phi\, c\, \beta (\beta-1) (S_i/A_i)^{\beta-1}] \tag{5.42}$$

and for c

$$F_i(S_i) = (S_i/A_i)^{\beta-1} / [1 + \phi\, \beta (S_i/A_i)^{\beta-1}] \tag{5.43}$$

The solution of these simultaneous non-linear equations can be obtained via the Newton-Raphson method described earlier.

Once the equilibrium supplied floorspace S_i^* is obtained from the above, the corresponding equilibrium number of firms is given as $N_i^* = S_i^*/u$. Following this, the corresponding equilibrium rents r_i^* are obtained from (5.28), the equilibrium wages w_i^* from (5.37) and the contact flows and agglomeration economies from (5.24) and (5.27) respectively.

Note that, given the equilibrium floorspace supplies S_i^*, it is then possible to maximize the total developer profit Π^* in (5.28) with respect to the developed land area A_i, which, given the land rents l_i, yields the unique optimal area A_i^* of released land as

$$A_i^* = S_i^* \sqrt[\beta]{[(\beta - 1)\, c / l_i]} \tag{5.44}$$

Of course, with the availability of a land market model, the released land A_i^* and the unit rent l_i^* would be co-determined.

5.3 Concluding Remarks

One of the major remaining challenges within the elementary model framework presented is to examine in depth the mathematical properties of the long run non-linear solution (5.29) for the equilibrium firm location pattern N_i^* when it is embedded into a logistic dynamic adjustment process, especially in light of recent work by Fujita and Thisse (2002). In particular, the properties of the alternative equilibria and associated bifurcations points need to be identified. For instance, some combinations of values of α, β, and h may lead to all knowledge activity concentrating at the CBD, whereas other values may yield either local clusters or a wide dispersion of activity. Some guidance on questions of uniqueness and existence is obtainable from the long run retail equilibrium analysis of Rijk and Vorst (1983). A further advance would be to model *gross* rather than *net* locational adjustments of the firms. This would allow re-location and entry of firms to be distinguished from exits, as well as allowing the 'free entry' assumption to be lifted.

A further goal is to model the demand for the outputs of the knowledge firms from government agencies and other private sector firms. In this respect, the revenue (R_i/N_i) generated per firm would become a proxy for *price* in this peculiar situation where the 'quantity' of knowledge is not readily identifiable.

6 Perfect and Imperfect Competition in Land/Housing Markets

6.1 Introduction

As discussed in Sect. 2.1.1, land and housing demand has been extensively modelled in economics as a key part of the field of 'urban economics' developed by Mills, Alonso and others. However, the modelling of the supply of land and associated housing has been studied much less. This has occurred probably for two main reasons (i) the difficulty of modelling the oligopolistic nature of these markets and (ii) the occurrence of several sets of agents in the process whose behaviour is interdependent. In this chapter, an attempt is made to meet these two challenges. It draws directly from joint research performed by the author and Professor Francisco Martinez, University of Chile, who has led this work.[1]

In the context of urban economics, demand and supply of locational choices can be described as two optimisation processes that provide the basic inputs to produce market equilibrium. Under the paradigm of rationality of consumers and suppliers, each optimisation process describes the expected behaviour of agents in the market.

The commodity traded in the market analysed here is an *urban* residential option, which is described by several attributes. Some are associated with the location of the land lots, and include the building and natural environment, as well as accessibility and attractiveness. Other attributes describe the land lot itself, including size and view, plus the dwelling, i.e. size, number of rooms, etc. Since Alonso (1964), urban economists have recognised that location attributes are not only associated with transport costs, but also with access to amenities related to the built environment. This property makes such a good very peculiar - it becomes quasi-unique. Attributes associated with the dwelling, on the other hand, do not introduce any peculiarity; they are produced and traded in the competitive building market. Fujita (1989) is a very useful reference here.

Alonso's theoretical framework of consumers' behaviour remains consistent with this economic condition of quasi-uniqueness by assuming that location options are traded in auctions; we can make the softer but perhaps more realistic assumption that agents have information on a *common value* at each location (see McAffe and McMillan 1987). This means that each location has a widely known

[1] As this is the only new work described from Chaps. 4 to 9 where I am not the 'senior author', I am most grateful to Francisco Martinez for his permission to include it here.

estimate of the value before the auction, but the final price is still defined by the rule of the highest bidder, who gets title to the property or the right to use this location. The theoretical discussion of this approach produced the bid-choice model (Martínez, 1992).

In contrast with normal competitive goods, where prices are defined by producers and expressed as a cost function, in this framework the price formation process has two components. The dwelling cost is a competitive component with a production function, whereas the location value is a monopolistic component resulting from capitalising the consumers preferences expressed by their willingness-to-pay.

Therefore, suppliers produce location-dwelling options subject to the conditions that the final price is defined by the best bid. Thus, their production problem can be described as choosing the *combination of location, land lots and dwellings that yields them the maximum profit.* Here we model this problem constrained by explicit planning regulations defining the available land for potential urban development during a specified time period.

6.2 The Economic Structure of the Models

6.2.1 Market Structure

The main feature of the model is that we decompose housing supply into the three-step production process shown in Fig. 6.1, where each step involves different producers or agents, extending the Ueda et. al. (1996) model. These steps describe the residential supply market as a production chain, starting with landowners, then land developers and finally housing developers. The model relates to new housing development rather than to redevelopment of existing housing.

Landowners (L). Agricultural land is made available for urban development. Agriculture landowners release a total of Q_i broadacres of land in each zone i to maximise their profit obtained from the price differential between the agricultural land price $\overline{p}_i$ and the release price p_i, paying a marketing cost $\overline{c}_i$ and subject to urban regulation policies.

Land Developers (LD). The land developer agent divides the released land in each zone into a number of land lots L_{ki} of different sizes q_k. This agent maximises the profit obtained from buying the land at a price p_i and selling an optimum number of lots of each size at a price p_{ki}, while paying out an exogenous marketing-management-development cost $\overline{c}_{ki}$. For computational simplicity, the land size is defined here as a discrete variable, with index k identifying the range of land lot sizes. Lots are discrete extracts from a continuous supply of land.

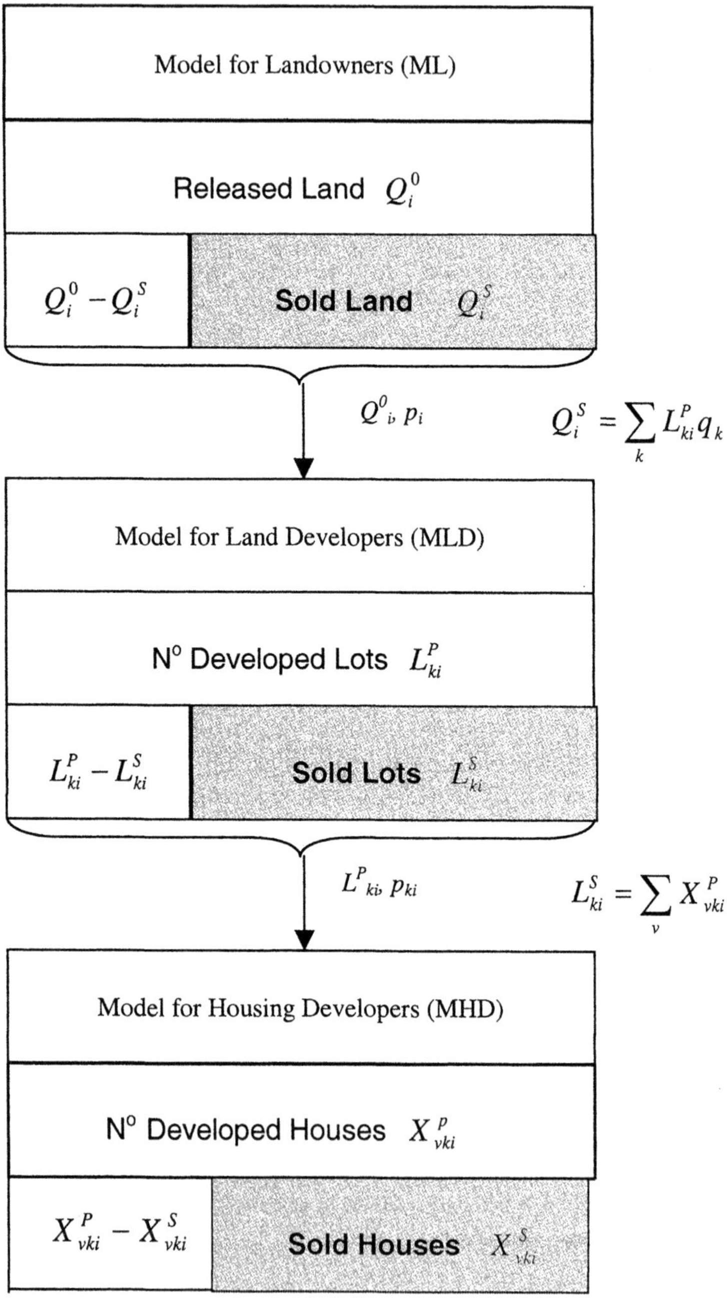

Fig. 6.1 Diagram of Model Structure

Housing Developers (HD). These agents build houses on each land lot. They decide the number of houses of type v to be built on each land lot X_{vki} in order to maximise the profit obtained from the price difference between buying the land p_{ki} and selling houses at prices r_{vki}, incurring exogenous building costs c_{vki}. Housing types are defined with regard to the building attributes, for example, detached, dual occupancy, etc., and may be extended to consider flats.

At this stage, it is useful to emphasise that the sequential multi-agent structure described above, and illustrated in Fig. 6.1, can be viewed as an archetype of many transformation processes in the economy. This interdependence structure is rarely identified explicitly in economic modelling. For example, iron ore may be extracted in a specific region, be transported to another region or country to be converted into steel, with the steel then being transported to other regions for use in construction. In principle, the above sequential scheme can also be applied to all such processes, especially if the number of successive transformative agents is not too large. There will be similar price and quantity signals flowing through these successive transformed products, as well as the presence of unsold stock in the longer run. If we then add the final step of modelling the demand by consumers, we have assembled the complete chain of events. In our housing/land model system, the rate of release of new land Q_i^0 for development would need to be coordinated with any unmet demand identified in the consumer demand analysis, introducing a feedback loop.

6.2.2 The Behavioural Model

A second relevant modelling element is that we assume, for each step, a free market where producers *maximise the profits* Π obtained from buying partially developed land, investing in some development, managing and marketing at a cost per unit area and selling at a profit. This behaviour is assumed to be aggregated – all agents are taken to be homogeneous and stochastic in applying a *maximum entropy* (or probabilistic) framework, extending the concepts introduced in Wilson (1970) and Roy and Lesse (1981). Additionally, observed information on total supply and demand is introduced to improve the fit of the model. This defines optimisation problems yielding demand and supply functions for each agent.

Economic Equilibrium. The important question about whether residential supply represents a perfect or imperfect competitive market is discussed here, deriving models for these two hypothetical cases in the context of comparative static equilibrium. In the first case, each agent behaves as a price taker, whilst in the second case agents can anticipate the consumers' demand function and extract their maximum willingness-to-pay. For this latter case, oligopolistic demand functions are obtained by inverting the competitive demand function in terms of supply price, which is then introduced into the supply function at the previous step in the market chain.

There is a strong theoretical argument, developed in the tradition of Alonso's (1964) urban economics, to support the thesis that landowners enjoy a non-competitive or oligopolistic land market. It is also plausible to assume that land and housing developers normally operate in a competitive market, because firms produce lots and houses anywhere, without having natural locational advantages. This is valid unless we consider that imperfect competition in the land market induces speculative behaviour in developers, who manage to partially capitalise locational advantages. Although there is the cost of extra complexity associated with modelling imperfect competition, the importance of this model is that it is thoroughly consistent with the bid-auction model of land-use briefly discussed above. This consistency is theoretically sound, because the argument that locational advantages justify the best bid rule in the land-use model is valid across the supply hierarchy. This makes the landowner the final holder of the monopolistic power associated with location, that is, the agent that capitalises, fully or partially, the locational willingness-to-pay of consumers.

Given this general economic framework we consider Walrasian equilibrium, which fulfils the market clearing condition, while complying with the condition that consumers are indifferent to all locations at the equilibrium price. Within this framework, it is possible to contemplate supply models aimed at forecasting the short and long run. In the urban land-use market however, the *short run* model would be defined as the case where total building stock is fixed and exogenous, which represents an economy having consumers only. Assuming an initial equilibrium, given by allocation of stock and prices, a new equilibrium is obtained under changes in consumers' preferences or marginal stock changes such as completion of ongoing development projects. This case is of no interest in our paper because supply is exogenous.

The *long run* case allows for supply adjustments to demand. The classical long run competitive equilibrium model assumes that all firms are identical – homogeneous - with the same technology and capital, which generates an equi-profit equilibrium. However, in the urban land market, goods are differentiated due to the quasi-unique characteristics of the land; thus landowners are not homogeneous as suppliers, because they hold monopoly power on spatially differentiated locational externalities, which yields differentiated profits. In this case, for the equi-profit condition to hold, it is required to assume some sort of market intervention. Since profit differentials are generated by locational externalities, then it sounds reasonable to consider that profit is maximised by each homogenous zone, independently of other zones. In other words, we assume monopoly within zones but oligopoly between zones.

The Walras condition may be specified under two optional assumptions. Firstly, we have the more classical assumption that supply exactly equals demand, which is modelled by specifying an *unconstrained* supply model. Secondly, we have a relaxed Walras condition that allows for some excess supply, asymptotically diminishing towards price equilibrium, modelled by *non-linear models with capacity constraints*, which generate production functions (denoted by superscript P) as well as functions for the stock S which is actually sold. We derive these functions for the landowners (Q_i^0, Q_i^S), land developers (L_{ki}^P, L_{ki}^S) and housing developers (X_{vki}^P, X_{vki}^S).

The Entropy Modelling Approach. Finally, it is important to relate the wider developments in entropy-type models discussed in Chap. 3, which are making them relevant in the context of economics, to this study in particular. These concepts are exprcssed in more detail in Roy (1997b). Typically, the only cost-related data present in the spatial entropy models of Wilson (1970) were transport costs. Thus, entropy models for determination of transport flows were interpreted as minimising the expected total cost of travel. However, in an important statement, Smith (1990) proposed that this 'behavioural' data in an entropy model should be encapsulated in constraints that are consistent with an appropriate model theory. Roy accepted this advice in using entropy to form *probabilistic* models demonstrating asymptotic convergence properties to the classical *deterministic* models of microeconomics, as demonstrated in Chap. 3. For instance, in the supply behaviour of the competitive firm (or an aggregate of 'identical' firms), the profit *objective* of the deterministic theory of the firm is replaced in the probabilistic analogue by a *constraint* on the profits *'observed'* at the base period, making the classical models special cases of the more general entropy models. Secondly, adapting ideas from Lesse (1982), it is possible to treat *Lagrange multipliers* of the estimated models as *parameters* in transformed versions for projection, allowing projected quantities to be endogenous and price-responsive, as demonstrated in Sect. 3.2.4. Thus, in the case mentioned above, the profits would emerge as *output* in the transformed models for the projection period.

In addition, when examining the issue of *available capacities* in modelling Walrasian equilibrium, advice was obtained from an early paper of Hotelling (1932). In a *regional* context, Hotelling demonstrated the rationale for generic *logistic* forms of regional supply and demand functions, which demonstrate *spill-over* effects into adjacent regions when capacity in any region is hard-pressed. Such *non-separable* functions can be obtained via entropy using a parallel argument, where units of available capacity are taken as *distinguishable* (or heterogeneous) in the entropy objective of the *constrained* model, acting somewhat like a compressing spring (repulsion effect) as the capacity limit in any zone is approached. This is discussed generally in Section 3.2.6.

Last, but not least, most probabilistic demand (and supply) functions arising from the entropy framework are analytically invertible. This opens the door to development of tractable model frameworks for imperfect competition.

The aim of this chapter is to present several models. In the following section, comparative static models of competitive markets are derived both for the constrained and unconstrained cases. The extension of these models to imperfect competition is presented in the next section. Finally, several models are formally presented in these two sections, leaving comments, interpretations and further analysis to the concluding section.

6.3 A Competitive Land Market

Although we have argued that the urban land market is normally imperfect, we shall start with models of a hypothetical competitive market for the sake of completeness and simplicity in the presentation. For the same reason, we first present the unconstrained model, so as to introduce later the extra complexity associated with the constrained model.

6.3.1 Unconstrained Models for Competitive Markets

The assumption in the unconstrained model is that supply from one step equals demand for the next agent in the chain.

Landowner Model (L/C/UNCNST)[2]. The entropy represents the number of ways that distinguishable units of land Q_i can be allocated into the different housing zones i. When this quantity is maximised under the necessary constraints, enforcing compliance with base period observations, we obtain the most likely quantities of land released. The entropy, given by

$$S^L = - \Sigma_i Q_i (\log Q_i - 1) \tag{6.1}$$

is maximised under the landowners' profit constraints given in terms of the selling price minus the cost of agricultural land and the marketing cost, given by

$$\pi^L = \Sigma_i Q_i (p_i - \bar{p}_i - \bar{c}_i) \tag{6.2}$$

which assumes landowning as an homogeneous industry. Additionally, total agricultural land potentially available for release for residential use is constrained by

$$\Sigma_i Q_i = Q^0 \tag{6.3}$$

With α_l the Lagrange multiplier on the profit constraint and λ_l that on the total quantity constraint, we obtain the Lagrangian function for each profit assumption

$$F^L = - \Sigma_i Q_i (\log Q_i - 1) + \alpha_l (\Sigma_i Q_i (p_i - \bar{p}_i - \bar{c}_i) - \pi_i) - \gamma_l (\Sigma_i Q_i - Q^0)$$

with first order conditions given by:

$$\partial F^L / \partial Q_i = - \log Q_i - \alpha_l (p_i - \bar{p}_i - \bar{c}_i) - \gamma_l = 0$$

[2] Model notation is: Agent (L, LD,HD) ; Competition (C- Perfect, IC-Imperfect) ; CNST and UNCNST (Constrained, Unconstrained)

This yields the landowners' profit maximising supply function:

$$Q_i = exp\,[\alpha_l(p_i - \bar{p}_i - \bar{c}_i) - \gamma_l] \tag{6.4}$$

In this model, α_l is calibrated by linear extrapolation to reproduce observed profit and/or supply data and γ_l is calibrated by the usual balancing factor fixed-point method so as to reproduce the total land constraint.

Alternatively, assuming that landowners are fully differentiable by zone due to location externalities, they maximise local zonal profit given by

$$\pi_i^L = \Sigma_i Q_i (p_i - \bar{p}_i - \bar{c}_i)$$

which leads to

$$F_i^L = - \Sigma_i Q_i (log\, Q_i - 1) + \alpha_{li} ([Q_i (p_i - \bar{p}_i - \bar{c}_i) - \pi_i^L] - \gamma_l (\Sigma_i Q_i - Q^0)$$

yielding the elegant result

$$Q_i = exp\,[\alpha_{li} (p_i - \bar{p}_i - \bar{c}_i) - \gamma_l] \tag{6.5}$$

Observed zonal profit levels to calibrate the α_{li} may be difficult to obtain.

Land Developer Model (LD/C/UNCNST). The total number of lots L_i^0 is split into L_{ki} lots of size q_k, yielding the following entropy:

$$S^{LD} = - \Sigma_{ki} L_{ki} (log\, L_{ki} - 1)$$

which is maximised subject to the aggregated profit constraint given as the lot selling price minus the buying price of the land in each lot minus associated costs:

$$\pi^{LD} = \Sigma_{ki} L_{ki} (p_{ki} - p_i q_k - \bar{c}_{ki})$$

and complementary quantity sum information from the observed land market introduced to the model, related to total lots L_i^0 in each zone i and the total lots L_k^0 of category k in the whole system:

$$L_i^0 = \Sigma_k L_{ki} \qquad L_k^0 = \Sigma_i L_{ki}$$

Thus, the first order conditions with Lagrangian multipliers α_2, γ_{i2} and η_k respectively, are:

$$\partial F^{LD} / \partial L_{ki} = - log\, L_{ki} + \alpha_2 (p_{ki} - p_i q_k - \bar{c}_{ki}) - \gamma_{i2} - \eta_k = 0$$

which yield the optimum values for the supplied lots:

$$L_{ki} = exp\ [\alpha_2 (p_{ki} - p_i\, q_k - \bar{c}_{ki}) - \gamma_{i2} - \eta_k] \tag{6.6}$$

Introducing the equilibrium condition between supplied lots and developed land

$$Q_i = \Sigma_k L_{ki}\, q_k$$

we can obtain the expression for the expected land price p_i^e for a unit of land via

$$p_i^e = \bar{p}_i + \bar{c}_i + (1/\alpha_1)\{ log\ \Sigma_k\ q_k\ exp\ [\alpha_2 (p_{ki} - p_i^e\, q_k - \bar{c}_{ki}) - \gamma_{i2} - \eta_k] - \gamma_1 \}$$

This equation expresses the expected price at equilibrium in a fixed-point format, that is $p_i^e = f(p_i^e)$, which cannot be solved analytically for p_i^e but has a numerical unique solution. Observe that numerical complexity is introduced by the change of units between areal land units and numbers of lots, which requires the transformation factor q_k.

Housing Developer Model (HD/C/UNCNST). In this case, land lots are further developed and the entropy expresses the number of ways in which lots can be enhanced with houses of different types ν in the form

$$S^{HD} = -\Sigma_{\nu ki} X_{\nu ki} (log\ X_{\nu ki} - 1)$$

which is maximised subject to the aggregated profit in terms of the unit 'rent' minus the cost of the lot itself minus building and associated costs:

$$\pi^{HD} = \Sigma_{\nu ki} X_{\nu ki} (r_{\nu ki} - p_{ki} - \bar{c}_{\nu ki})$$

and the conditions from observed quantity information, in terms of the total number of houses X_i^0 built in each zone and the number $X_{\nu k}{}^0$ of houses built of type ν on lots of size category k:

$$\Sigma_{\nu k} X_{\nu ki} = X_i^0 \quad ; \quad \Sigma_i X_{\nu ki} = X_{\nu k}{}^0$$

It follows that:

$$\partial F^{HD} / \partial X_{\nu ki} = -log\ X_{\nu ki} + \alpha_3 (r_{\nu ki} - p_{ki} - \bar{c}_{\nu ki}) - \gamma_{i3} - \eta_{\nu k} = 0$$

yielding the housing supply function as:

$$X_{\nu ki} = exp\ [\alpha_3 (r_{\nu ki} - p_{ki} - \bar{c}_{\nu ki}) - \gamma_{i3} - \eta_{\nu k}] \tag{6.7}$$

Introducing the supply/demand consistency equation :

$$L_{ki} = \Sigma_v X_{vki}$$

we can obtain the expression for lot prices at equilibrium p_{ki}^e, given by:

$$p_{ki}^e = 1/(\alpha_2 + \alpha_3)\{\alpha_2 (p_i q_k + \bar{c}_{ki}) - \gamma_{i2} - \eta_k + log\Sigma_v exp[\alpha_3 (r_{vki} - \bar{c}_{vki}) - \gamma_{i3} - \eta_{vk}]\}$$

In this case, the price function is explicit and depends on land prices as the production factor, and on selling prices of house options using the lot size q_k.

6.3.2 Constrained Models for Competitive Markets

These models assume that the total supplied units may be in excess of the actual demand, which requires some extensions of the previous unconstrained models in this direction. In the specification of excess supply one may assume a linear or non-linear decrease of stock. In this latter and perhaps more interesting case, the entropy function may be specified by an extended model recognising distinguishable capacity units, where the *slope* of the supply function decreases in a *logistic* form with price, as production approaches capacity (Hotelling, 1932). Such forms and their relevance are discussed in detail in Sect. 3.2.6, where a *distribution* across the 'utility' of the units in each zone is implied.

Landowner Model (L/C/CNST). From a total of Q_i^0 broadacres of building land released for development in the base year in zone i, Q_i^S is sold. The entropy for zone i now represents the number of ways that the units of supplied land and unsold land can be allocated from within the distinguishable units of capacity, written as:

$$S_i^L = log[Q_i^0! / (Q_i^0 - Q_i^S)! \, Q_i^S!]$$

which yields the following objective:

$$S^L = -\Sigma_i (Q_i^0 - Q_i^S) [log(Q_i^0 - Q_i^S) - 1] - \Sigma_i Q_i^S [log Q_i^S - 1]$$

The entropy is maximised under the following constraints:

$$\Sigma_i Q_i^S = Q^0 \quad ; \quad \pi^L = [Q_i^S (p_i - \bar{c}_i) - Q_i^S \bar{p}_i - (Q_i^0 - Q_i^S) I \bar{p}_i]$$

where I represents the interest rate which generates the cost of holding the unsold stock $(Q_i^0 - Q_i^S)$ in each zone i. With γ_1 and α_1 as the respective Lagrange multipliers, we obtain

$$Q_i^S = Q_i^0 \exp[\alpha_1(p_i - \bar{p}_i I - \bar{c}_i) - \gamma_1] / \{1 + \exp[\alpha_1(p_i - \bar{p}_i I - \bar{c}_i) - \gamma_1]\} \quad (6.8)$$

This may be calibrated by linear extrapolation for α_1 and via fixed-point problem for γ_1.

Land Developer Model (LD/C/CNST). The total number of lots L_i^0 is subdivided by lot size into the number of produced lots L_{ki}^P and sold lots L_{ki}^S. The entropy is:

$$S^{LD} = [\pi_i L_i^0! / (\pi_k L_{ki}^P!)] \; [\pi_{ik} L_{ki}^P! / L_{ki}^S! (L_{ki}^P - L_{ki}^S)!]$$

This reduces to

$$S^{LD} = -\Sigma_{ki} L_{ki}^S [\log L_{ki}^S - 1] - \Sigma_{ki} (L_{ki}^P - L_{ki}^S) [\log (L_{ki}^P - L_{ki}^S) - 1]$$

being maximised under the following total quantity constraints:

$$\Sigma_k L_{ki}^P = L_i^{PO} \quad ; \quad \Sigma_i L_{ki}^P = L_k^{PO}$$

$$\Sigma_k L_{ki}^S = L_i^{SO} \quad ; \quad \Sigma_i L_{ki}^S = L_k^{SO}$$

plus the usual profit constraint including holding costs:

$$\pi^{LD} = \Sigma_{ki} (L_{ki}^S p_{ki} - L_{ki}^P p_i q_k - L_{ki}^P \bar{c}_{ki}) - \Sigma_{ki} (L_{ki}^P - L_{ki}^S) I p_i q_k$$

If these are associated with Lagrange multipliers ϕ_i, χ_k, γ_{i2}, η_k and α_2 respectively, we finally are able to write:

$$L_{ki}^P = \exp - [\alpha_2(p_i q_k + \bar{c}_{ki} + I p_i q_k) + \phi_i + \chi_k] \{1 + \exp[\alpha_2(p_{ki} + I p_i q_k) - \gamma_{i2} - \eta_k]\} \quad (6.9)$$

as well as the relation

$$L_{ki}^S = \exp - [\alpha_2(p_i q_k + \bar{c}_{ki} + I p_i q_k) + \phi_i + \chi_k] \,.\, \exp[\alpha_2(p_{ki} + I p_i q_k) - \gamma_{i2} - \eta_k] \quad (6.10)$$

If we introduce the consistency condition $\Sigma_k q_k L_{ki}^P = Q_i^S$, using (6.8) and (6.9), we obtain the following equation for the equilibrium land price:

$$Q_i^0 \frac{\exp[\alpha_1(p_i^e - \bar{p}_i I - \bar{c}_i) - \gamma_1]}{\{1 + \exp[\alpha_1(p_i^e - \bar{p}_i I - \bar{c}_i) - \gamma_1]\}} = \frac{\Sigma_k [q_k \{1 + \exp[\alpha_2(p_{ki} + I p_i^e q_k) - \gamma_{i2} - \eta_k]\}.}{\exp - \{\alpha_2(p_i^e q_k + \bar{c}_{ki} + I p_i^e q_k) + \phi_i + \chi_k\}]}$$

Housing Developer Model (HD/C/CNST). Agents allocate to the supply L_{ki}^S in each zone i of distinguishable lots for housing development X_{vki}^P houses of type v, of which X_{vki}^S get sold. Then, the system entropy becomes

$$S^{HD} = log\{ [\pi_{ki} L_{ki}^P! / (L_{ki}^P - \Sigma_v X_{vki}^P)!\ \pi_v X_{vki}^P!] [X_{vki}^P! / (X_{vki}^P - X_{vki}^S)!\ X_{vki}^S!]\}$$

With the usual approximations, S is written after some cancellation as:

$$S^{HD} = - \Sigma_{ki} (L_{ki}^P - \Sigma_v X_{vki}^P) [log (L_{ki}^P - \Sigma_v X_{vki}^P) - 1] - \Sigma_{vki} (X_{vki}^P - X_{vki}^S) [log (X_{vki}^P - X_{vki}^S) - 1] - \Sigma_{vki} X_{vki}^S [log X_{vki}^S - 1]$$

This is maximized under the following constraints

$$\Sigma_i X_{vki}^S = X_{vk}^{SO} \quad ; \quad \Sigma_k X_{vki}^S = X_{vi}^{SO} \quad ; \quad \Sigma_v X_{vki}^S = X_{ki}^{SO}$$

$$\Sigma_{vk} X_{vki}^P = X_i^0 \quad ; \quad \Sigma_i X_{vki}^P = X_{vk}^0$$

$$\pi^{HD} = \Sigma_{vki} [X_{vki}^S r_{vki} - X_{vki}^P (p_{ki} + c_{vki}) - (X_{vki}^P - X_{vki}^S) I p_{ki}]$$

attached to the objective with Lagrange multipliers λ_{vk}, ϕ_{vi}, ψ_{ki}, γ_{i3}, η_{vk} and α_3 respectively. Upon differentiation with respect to X_{vki}^P and X_{vki}^S, equating to zero and performing summations to remove X_{vki}^S from the right-hand side, we finally obtain the production of housing

$$X_{vki}^P = L_{ki}^P exp - \rho_{vki}^P [1 + exp\ \rho_{vki}^S] / \{1 + \Sigma_v exp - \rho_{vki}^P [1 + exp\ \rho_{vki}^S]\} \quad (6.11)$$

with the amount sold of each type on each lot category being:

$$X_{vki}^S = L_{ki}^P exp (\rho_{vki}^S - \rho_{vki}^P) / \{1 + exp - \rho_{vki}^P [1 + \rho_{vki}^S]\} \quad (6.12)$$

with the ρ symbols defined as

$$\rho_{vki}^P = \alpha_3 (p_{ki} - c_{vki} + I p_{ki}) - \gamma_{i3} - \eta_{vk}$$

$$\rho_{vki}^S = \alpha_3 (r_{vki} + I p_{ki}) + \lambda_{vk} + \phi_{vi} + \psi_{ki}$$

Introducing the equilibrium condition between sold lots and those transformed to housing lots, given by $L_{ki}^S = \Sigma_v X_{vki}^P$, we obtain:

$$L_{ki}^S = \Sigma_v X_{vki}^P = A_{ki} / (1 + A_{ki}) \quad (6.13)$$

with $A_{ki} = \Sigma_v exp - \rho_{ki}^P (1 + exp\ \rho_{ki}^S)$, and using equations (6.9) and (6.10):

$$L_{ki}^P = B_{ki} (1 + C_{ki}) \quad ; \quad L_{ki}^S = B_{ki} C_{ki} \quad (6.14)$$

with $B_{ki} = exp- [\alpha_2 (p_i q_k + \bar{c}_{ki} + I p_i q_k) + \phi_i + \chi_k]$ and $C_{ki} = exp [\alpha_2 (p_{ki} + I p_i q_k) - \gamma_{i2} - \eta_k)]$. Upon substituting (6.14) into (6.13), we obtain a market equilibrium condition $A_{ki} = C_{ki}$. To extract p_{ki} it is necessary to assume no holding costs ($I=0$), which yields:

$$p_{ki}^{e} = (1/\alpha_2 + \alpha_3)\{\log\{\Sigma_v \exp[\alpha_3 c_{vki} - \gamma_{i3} - \eta_{vk}] + \exp[\alpha_3 (r_{vki} + c_{vki}) - \gamma_{i3} - \eta_{vk} - \lambda_{vk} - \phi_{vi} - \psi_{ki})]\} - \gamma_{i2} - \eta_k\} \quad (6.15)$$

which is the expected equilibrium price for lots k in zone i. It is worth noting that this price is expressed in terms of model parameters only, except for the house price or rent r_{vki}, which make these lot prices dependent on the consumer and supply equilibrium in the residential housing market.

6.3.3 Hotelling's Lemma

This lemma, also called the derivative property of profit functions, states that the firm's net supply function is equal to the price derivative of the profit function. In other words, it should be possible to obtain the supply function from the profit function; in fact this is the approach used by Ueda *et al.* (1996) to derive supply and demand functions. In this section, we shall prove the lemma for the probabilistic land developer supply function of the constrained model in (6.10). It was already proved in Sect. 3.2.9 for a typical producer/consumer problem.

Let us establish $(1/\alpha_2)$ times the Legèndre transform (Lesse, 1982) of the total original entropy objective F, converting it into the 'equivalent' *unconstrained* objective, identified as an expected profit. This quotient scaling factor is required because of the presence of the Lagrange multiplier α_2 in front of the deterministic profit relation in the equivalent unconstrained Lagrangian. This means that we are effectively maximizing the deterministic profit plus a 'dispersion term' coming from the equivalently scaled entropy objective itself. Let us then substitute the solutions into this Lagrangian as follows:

$$Z = (1/\alpha_2)\ \{\ \Sigma_i L_i^0 [\log L_i^0 - 1] + \Sigma_{ik} L_{ki}^S [-\phi_i - \chi_k - \lambda_{i2} q_k + \alpha_2 \bar{c}_{ki} + \alpha_2 p_i q_k + \alpha_2 p_i I q_k - \gamma_{i2} - \eta_k + \alpha_2 p_{ki} - \alpha_2 p_i I q_k + 1] + \Sigma_{ik} (L_{ki}^P - L_{ki}^S) + \Sigma_{ik} L_{ki}^P [-\phi_i - \chi_k + \alpha_2 \bar{c}_{ki} + \alpha_2 p_i I q_k + \alpha_2 p_i q_k] + \Sigma_{ik} L_{ki}^P [-\phi_i - \chi_k + \alpha_2 \bar{c}_{ki} + \alpha_2 p_i q_k + \alpha_2 q_k I p_i] + \Sigma_{ik} (\phi_i L_{ki}^P + \chi_k L_{ki}^P) + \Sigma_{ik} (\gamma_{i2} L_{ki}^S + \eta_k L_{ki}^S) + \alpha_2 \Sigma_{ik} (L_{ki}^S p_{ki} - L_{ki}^P \bar{c}_{ki} - L_{ki}^P p_i q_k - (L_{ki}^P - L_{ki}^S) q_k I p_i) + K\}$$

with K a constant. After considerable cancellation, we finally obtain:

$$Z = (1/\alpha_2)\{\Sigma_i L_i^0 (\log L_i^0 - 1) + (\Sigma_{ik} L_{ki}^P) + K\}$$

Now, substituting from (6.10), we can prove Hotelling's Lemma as follows:

$$\partial Z / \partial p_{ki} = (1/\alpha_2)[\alpha_2 L_{ki}^P] = L_{ki}^P \qquad \text{Q.E.D.}$$

where an increase in the output price increases profit. Thus, our probabilistic result maintains this fundamental property of the classical deterministic theory.

6.4 Imperfect Competition

Within each zone, sets of land developers are taken to supply lots, and landowners land, in anticipation of the willingness-to-pay for their outputs by the demand agents. In this case, their expected behaviour is to maximise their oligopolistic profit. For the housing developers, this is implicit in the above model when rents (r_{vki}) represent the expected maximum bid value, which is anticipated by these agents. Analytically, this means that in the calculation of first order conditions, prices should be differentiated as appropriate functions of quantities. Imperfect competition is represented here in terms of Cournot oligopoly (Intriligator, 1971), where each 'agent' just considers his opponents' current state, rather than trying to predict how each opponent will react to his own decisions. On the other hand, in common with all models of imperfect competition, he is assumed to know how demand will adjust to his supply decisions, implying that the inverse demand functions replace input prices within the profit functions of the oligopolists. These then become differentiable with respect to the supply quantities.

6.4.1 Unconstrained Models for Imperfect Competition

Landowner Model (L/IC/UNCNST). In contrast to the competitive case, here land prices are assumed as dependent on supply L_{ki}, then:

$$\partial F / \partial Q_i = - \log Q_i - \alpha_1 (p_i + c_i) - \gamma_1 - \alpha_1 Q_i \, \partial(p_i^*) / \partial Q_i$$

To obtain the oligopolistic price p_i^*, we should invert the aggregated demand function L_{ki} in p_i [using eq (6.6)]. Take $Q_i = \Sigma_v q_k L_{ki}$ given by:

$$Q_i = \Sigma_k q_k \exp [\alpha_2 (p_{ki} - p_i q_k - \bar{c}_{ki}) - \gamma_{i2} - \eta_k]$$

Differentiating implicitly, we obtain:

$$\partial Q_i / \partial p_i = - \alpha_2 \Sigma_k q_k^2 \exp [\alpha_2 (p_{ki} \; p_i q_k - \bar{c}_{ki}) - \gamma_{i2} - \eta_k]$$

which yields:

$$Q_i [\partial p_i / \partial Q_i] = - [\Sigma_k q_k L_{ki}] / [\alpha_2 \Sigma_k q_k^2 L_{ki}]$$

Also, we have from equation (6.6):

$$p_i^* = (1 / \alpha_2 q_k) [\alpha_2 (p_{ki} - \bar{c}_{ki}) - \log L_{ki} - \gamma_{i2} - \eta_k] \tag{6.16}$$

Then replacing the above expressions in $\partial F / \partial Q_i = 0$, we obtain a fixed-point equation for L_{ki}. With the solution for L_{ki}, we can obtain the value for Q_i as the land supply quantity for imperfect competition.

Land Developer Model (LD/IC/UNCNST). Similarly, we assume lot prices as being dependent on supply, yielding:

$$\partial F / \partial L_{ki} = - \log L_{ki} - \gamma_{i2} - \eta_k - \alpha_2 (p_i q_k + \bar{c}_{ki}) + \alpha_2 [p_{ki}^* + (\partial p_{ki}^* / \partial L_{ki})]$$

In order to derive the oligopolistic price p_{ki}^*, we invert the aggregate demand function $\Sigma_v X_{vki}$ in p_{ki} [using equation (6.7)] and taking $L_{ki} = \Sigma_v X_{vki}$ we obtain

$$p_{ki}^* = (1 / \alpha_3) \log \{ [\Sigma_v \exp \{ \alpha_3 (r_{vki} - \bar{c}_{vki}) - \gamma_{i3} - \eta_{vk}\}] / L_{ki} \}$$

which yields $L_{ki} (\partial p_{ki}^* / \partial L_{ki}) = - (1 / \alpha_3)$.

Then the oligopolistic supply function L_{ki}^* is

$$L_{ki}^* = \exp [\alpha_2 (p_{ki}^* - p_i q_k - \bar{c}_{ki} - 1/\alpha_3) - \gamma_{i2} - \eta_k]$$

and replacing p_{ki}^* we obtain :

$$L_{ki}^* = \{ \exp - [\alpha_2 (p_i q_k + \bar{c}_{ki} + 1/\alpha_3) + \gamma_{i2} + \eta_k] \}^{\alpha_3/(\alpha_3 - \alpha_2)} \cdot \rho_{ki}^{\alpha_2/(\alpha_3 - \alpha_2)} \tag{6.17}$$

where $\rho_{ki} = \Sigma_v \exp [\alpha_3 (r_{vki} - \bar{c}_{vki} - \gamma_{i3} - \eta_{vk}]$.

It is of interest to use this model to compare (6.6) for perfect competition with the corresponding (6.17) for the imperfect case. The main difference, in addition to the extra - $(1/\alpha_3)$ term, is that a power of a 'composite' price ρ_{ki} replaces the market price p_{ki}. This has a relationship with the occurrence of 'inclusive value' terms within *nested* models in discrete choice theory, indicated in the log-sum price terms p_{ki}^*. The result is that the pure sequential nature of our three sets of perfect competition models is overcome by the 'looking forward' by the land developers to the demand for the different types of housing v.

Housing Developer Model (HD/IC/UNCNST). Additionally, equations (6.11) and (6.12) naturally give the supply model for the housing developer under imperfect competition, because for the whole three steps of the supply model, residential demand prices are assumed exogenous.

6.4.2 Constrained Models for Imperfect Competition

Landowner Model (L/IC/CNST). Here the demand function $Q_i^S = \Sigma_k q_k L_{ki}^P$ of the land developers, with L_{ki}^P expressed via (6.9), is not analytically invertible to give p_i as an explicit function of Q_i^S. However, it is possible to use implicit differentiation to evaluate $(\partial p_i / \partial Q_i^S)$ analytically. This enables the landownner supply function (6.8) to take the form

$$Q_i^S = Q_i^0 \exp[\alpha_1 (p_i + Q_i^S \partial p_i / \partial Q_i^S - \bar{p}_i I - \bar{c}_i) - \gamma_1] / \{1 + \exp[\alpha_1 (p_i + Q_i^S \partial p_i / \partial Q_i^S - \bar{p}_i I - \bar{c}_i) - \gamma_1]\} \quad (6.18)$$

We note here that, as distinct from the competitive case in (6.8), the prices p_i are not treated as exogenous, but as functions of the quantities Q_i^S. This means that (6.18) must be solved iteratively, with p_i being obtained for the current value of Q_i^S at each iteration by solution of $Q_i^S = \Sigma_k q_k L_{ki}^P$ numerically, where L_{ki}^P is expressed in terms of p_i via (6.9). Although the procedure would be more direct if (6.9) were analytically invertible, the numerical procedure is still plausible.

Land Developer Model (LD/IC/CNST). First order conditions of the optimization problem LD/C/CNST with variable prices are:

$$\partial F / \partial L_{ki}^S = -\log L_{ki}^S + \log(L_{ki}^P - L_{ki}^S) - \gamma_{i2} - \eta_k + \alpha_2 (I p_i q_k) + \alpha_2 \partial(L_{ki}^S p_{ki}^*) / \partial L_{ki}^S$$

Analogously to the unconstrained case, here we obtain the aggregate housing developer demand function for land lots $L_{ki}^S = \Sigma_v X_{vki}^P$ from equation (6.11). This can be seen as non-analytically invertible if we retain the holding cost term $(I p_{ki})$. However, this will be quite small for cases of fairly low interest rates and low unused capacity. Thus, neglecting this term, the inversion comes directly from (6.11) as

$$p_{ki}^* = (1/\alpha_3) \{ \log [(L_{ki}^P - L_{ki}^S) G_{ki} / L_{ki}^S]\} \quad (6.19)$$

with $G_{ki} = \Sigma_v \exp \alpha_3 c_{vki} (1 + \rho_{vki}^S)$, which does not depend on p_{ki}^*. Then,

$$L_{ki}^S (\partial p_{ki}^* / \partial L_{ki}^S) = -(1/\alpha_3) L_{ki}^P / (L_{ki}^P - L_{ki}^S)$$

and the supply function for land developers (including holding costs) is:

$$(L_{ki}^{S*})^{[1+(\alpha_2/\alpha_3)]} = (L_{ki}^P - L_{ki}^{S*})^{[1+(\alpha_2/\alpha_3)]} \exp - \{ \gamma_{i2} + \eta_k - \alpha_2 I p_i q_k + (\alpha_2/\alpha_3) L_{ki}^P / (L_{ki}^P - L_{ki}^S)\} \quad (6.20)$$

If we use successive substitution, this equation becomes a special type of fixed-point problem whose convergence needs further investigation.

6.5 Final Comments

Each model may potentially be enriched with other informational constraints, either to enhance those introduced above or to incorporate planning regulations. This 'natural' way of handling constraints is a virtue of the probabilistic entropy maximization approach. The Lagrange multipliers on these constraints are then transformed into parameters for projection, as shown in Sec 3.2.4, allowing the associated quantities to be endogenous output.

A more complete analysis of parameter calibration is required, as we have only mentioned methods that have proved to be successfully applied in similar models. However, we foresee that the more complex expressions presented here may require an extension of the usual linear extrapolation and fixed-point methods, such as the Newton-Raphson approach.

Imperfect competition models generate supply functions that represent fixed-point problems. The question of whether or not the various solutions exist or are unique requires further research. Although we may expect a unique equilibrium between each pair of markets, the question still remains open as to whether or not we have a unique result over the complete set of linked markets. Each model of imperfect competition may be interpreted as if the agents maximise their profit in a stochastic framework, which follows directly from the invariant Legèndre transform, where the transformed expected profits expression substitutes for the optimising profit objective of the corresponding deterministic model. Based on this interpretation, the supply model has an orthodox economic interpretation.

The supply models developed here contribute to land-use models describing the behaviour of the supply side, which can be used in a demand-supply equilibrium process. For example, they can be directly used as the supply functions of the bid-choice model applied to Santiago, called MUSSA (Martínez and Donoso, 1995). Under this consistent economic model, the land-use market equilibrium can be properly characterised.

These supply models describe the transmission mechanism of the landowner's monopolist power, associated with location advantages, that is, access and location externalities, through the chain from land developers to the final consumer. Conversely, they describe how the consumer's valuation of these locational attributes is capitalised by landowners and intermediate developers. Appropriate calibration of parameters defines the proportion of the power captured by each agent. In a wider context, this mechanism applies to the supply process of any goods with a quasi-unique characteristic.

We have developed a set of models under unconstrained and constrained conditions. Upon their application to real cities, specific market conditions should be assessed. From general analysis and experience it seems plausible to consider imperfect competition and test whether the constrained or unconstrained models perform best. The new models will then represent theoretically well-founded means for further understanding the complex links between urban land development and transport investment. In addition, similar principles to the above may be applied to analyse the influence of sets of interdependent product transformation sequences during the provision of final demand entities.

Part III
Some Interregional Contributions

7 Trade Under Changing Networks and Technology

In Sect. 2.2.1, classical national, interregional and multi-regional input-output analysis is described. Although transport itself can be defined as a sector in the input-output tables, linear input-output analysis, or even the non-linear CGE (Computable General Equilibrium) analysis, provide quite 'blunt instruments' for investigations of the influences of transport network changes on the patterns of trade and associated regional output. Analysis is further complicated if changes of technology are occurring at the same times as the transport network changes or changes in transport pricing.

In the following, some partly new approaches to assessing trade patterns by both multi-regional and interregional input-output analysis are described. If we are primarily interested in assessing the regional economic influences of major transport infrastructure developments, the associated 'technology' can be included by inducing the flows to remain consistent with fundamental input-output relations. As regional outputs should be endogenous in the projection period, it is important to provide guidance in the determination of the intermediate and final demand flows for our short run models via the inclusion of long run sector capacities. Also, we examine possible enhancements of the entropy approach via a more general procedure developed from information theory. In addition, one should consider uncertainty in the basic production functions driving the input-output approach. On the other hand, observed changes in output, final demand and transport networks and prices should enable the resulting changes in the input-output coefficients themselves to be inferred. Both these problem classes are addressed below. Finally, we examine the influence of data availability on the appropriate level of detail that the models can reproduce, as well as on the flow definitions that correspond to measurable flows. For example, input-output coefficients are normally just available for each region of production (that is, multi-regional coefficients), and not as full interregional coefficients. In addition, commodity flows are usually just obtainable between regions, without any indication of the sector to which each commodity is to be provided as intermediate input or final demand in the destination regions. These features make the analysis quite challenging. The approaches shown below to handle these problems lead to the identification of some adjustments to earlier methods. We examine both interregional and multi-regional analysis with interregional or multi-regional data.

7.1 Trade Under a Changing Transport Network with Given Technology

Balance relations, as inferred from the 'supply and demand pool' assumptions in the multi-regional model of Leontief and Strout (1963), were applied by Wilson (1970) as technological constraints on a probabilistic multi-regional model of the flow of commodities. The model, based on entropy maximization, was also constrained to reproduce the observed average travel cost per unit of commodity shipped, integrating the technology and gravity effects in the flow determination. In this section, a multi-regional approach is developed in terms of expected data availability. Whereas the demand pool assumption of Leontief-Strout (LS) is made directly in our approach, their supply pool assumption is related purely to flow data availability. The first step is the provision of an interregional set of usage relations of regional output which explicitly distinguish between the intermediate flows and the flows directly to final demand in other regions, as well as the external export and import flows within the interregional system. This explicit identification of such extra flows turns out to be critical in inducing our approach to mesh with measurable flow data.[1] The resulting usage equations yield a significant adjustment to conventional interregional analysis. Then we assume the availability of just multi-regional data to evaluate either aggregated multi-regional flows or interregional flows. With the basic demand-driven I-O model relating to *outflows* from the output of each sector in each region, we derive balance relations in terms of *inflows*, with the flows now defined as aggregated over each sector of supply, including final demand. Despite our more detailed derivation, these balance relations for the multi-regional case turn out to be similar in spirit to the LS relations. The next step is to enhance the entropy maximization framework of Wilson (1970) by the revised balance relations, plus an allowance for regional interdependencies, by introducing logistic constraints on regional output capacity. Further extensions of the approach lead to the generation of probabilistic supply functions as tools within a potential CGE analysis. This option requires the introduction of prices, permitting a profit constraint to replace the simple transport cost constraint of the earlier model.

The multi-regional LS input-output approach built on the Chenery-Moses model introduced by Chenery (1953). The resulting model turns out to be much more tractable than a full interregional model, and has been widely applied. Leontief and Strout used an independent gravity type model to estimate the resulting trade flows, as a substitute for the provision of trade coefficients in models from the Chenery-Moses tradition. Wilson (1970) provided a more integrated procedure, setting up a conventional entropy framework to determine the commodity flows enhanced by the Leontief-Strout input-output balance relations introduced as constraints. In other words, the flows were co-determined by the transport cost

[1] I am indebted to an anonymous reviewer for this important insight for internal flows, as well as to Suahasil Nazara, Regional Economics Applications Laboratory (REAL), for suggesting the inclusion of the external import and export flows in the analysis.

information together with the technological information embodied in the Leontief-Strout representation of input-output. Another enhancement was to replace the entropy framework by one from information theory based on historical trade patterns (Snickars and Weibull, 1977), yielding models such as in Batten (1983).

In Roy (1997b), the introduction of uncertainty into the producer profit terms allowed prices to be retained explicitly in a regional input-output framework. Although the retention of prices is desirable from a theoretical viewpoint, such price information at the regional level is often extremely difficult to obtain. In fact, the whole of regional input-output analysis is plagued by a paucity of survey data. So, in this section, the main emphasis is to base the multi-regional model on just the survey data obtainable without a major effort. In fact, the LS demand pool assumption is entirely equivalent to a condition based on data availability, that is, that input-output coefficients are just defined at the region of production, independently of the mix of the inputs from other regions. However, their supply pool assumption is replaced here by the condition that the flows of commodities between regions are only known in terms of the commodity (viz. sector) being shipped, independently of the sector or sectors to which they are being supplied as intermediate inputs or final demand. Slightly revised balance relations are derived here, compared with the conventional LS relations and introduced into an entropy maximization framework, which may be set up to determine either interregional or multi-regional flows.

In any short run model of regional supply, it is desirable to include output capacity constraints. Of course, an obvious way to achieve this is to introduce ≤ inequality constraints on regional output capacity for each sector. However, such constraints, when inactive, maintain separability and have no influence whatsoever on the flows - they only factor into the analysis once they become active. Intuitively, this is not very plausible. In most sectors, a 'vintage' distribution over regional capacity exists, and when the capacity limits of a sector are being hard pressed in a certain region, one would expect some spillovers into adjacent regions. In fact, the generic logistic forms of regional supply functions motivated by Hotelling (1932) and discussed by Johansson (1991) demonstrate this property. In this paper, an additional entropy term recognising heterogeneity within the available capacity is shown to yield a logistic supply function. This provides a significant enhancement to the Wilson framework, especially for sectors within regions where there is a pronounced vintage structure of producer firms. Of course, if regional price information were available, our transport cost constraint should be replaced by a short run constraint on profits, allowing our result to truly be interpreted as a supply function. In addition, by use of the special information theory method of Roy (1987), the model not only perfectly reproduces the base period flows (as for conventional information theory models), but is simultaneously responsive to future changes in the input-output coefficients, regional output capacities, freight prices and the transport network, with the technological balance relations being imposed anew in the projection time period. Finally, these results are generalized by the introduction of explicit market prices, and compared with the formulation in Roy (1997b). Short run profit constraints replace the simple transport cost constraints, flows are defined in quantity rather than money units and the supply to final demand is endogenous. The resulting flow relations then represent logistic input demand functions.

7.1.1 An Enhanced Deterministic Interregional Model

In the course of introduction of our multi-regional approach, it is important to proceed from a 'full' interregional analysis, from which the argument of data availability provides the required 'pruning'. Also, we want to track the influence on the analysis of the inclusion of interregional flows direct to final demand, as well as external export and import flows, on the overall interregional system. In addition to having the conventional interregional flows x_{ij}^{rs} of sector i from region r as internal intermediate inputs to sector j in region s, we also include the corresponding intermediate flows i_{ij}^{rs} to region s from external imports into region r. Assuming a common technology, the interregional input-output coefficients a_{ij}^{rs} are now defined for the sum of intermediate inputs as

$$a_{ij}^{rs} = (x_{ij}^{rs} + i_{ij}^{rs}) / X_j^s \tag{7.1}$$

expressing the quotient of the total inputs $(x_{ij}^{rs} + i_{ij}^{rs})$ of sector i from region r into sector j in region s over the total output X_j^s of sector j in region s. Then, the *usage* of the output X_i^r of sector i in region r is expressed as that going to supply intermediate inputs of all sectors into all regions plus the amounts going to final demand and to external exports, both in its own and *other* regions. Defining y_I^{rs} as the flow of final demand of sector i from producing region r to consumption region s and e_i^{rs} as the flow of sector i from producing region r for external export out of region s, the usage relations then become

$$X_i^r = \Sigma_{js}\, x_{ij}^{rs} + \Sigma_s\, y_i^{rs} + \Sigma_s\, e_i^{rs} \tag{7.2}$$

where within-region usage is denoted by indices (r,r). Now we should express the final demand Y_i^r *satisfied* in region r in terms of our new flow quantities y_i^{sr} and the external import flows i_i^{sr} directly into final demand, yielding

$$Y_i^r = \Sigma_s\, y_i^{sr} + \Sigma_s\, i_i^{sr} \tag{7.3}$$

The final step is to solve (7.2) for the outputs X_i^r. We need to define internal final demand *exports* Z_i^r (including consumption in its own region), the goods $\bar{E}_i^r$ of sector i produced at region r and exported abroad from all regions s, the imports $\bar{I}_i^r$ of sector i from all regions s to be utilised at region r and the imports $\tilde{I}_i^r$ of sector i into region r for shipment to all regions s for *intermediate* inputs, as

$$Z_i^r = \Sigma_s\, y_i^{rs} \;;\quad \bar{E}_i^r = \Sigma_s\, e_i^{rs} \;;\quad \bar{I}_i^r = \Sigma_s\, i_i^{sr} + \Sigma_{js}\, i_{ij}^{sr} \;;\quad \tilde{I}_i^r = \Sigma_{js}\, i_{ij}^{rs} \tag{7.4}$$

Note that, the usual total exports abroad E_i^r of sector i out of region r and the total imports I_i^r of sector i from abroad into region r are given as

$$E_i^r = \Sigma_s e_i^{sr} \quad ; \quad I_i^r = \Sigma_{js} i_{ij}^{rs} + \Sigma_s i_i^{rs} \tag{7.5}$$

which differ from the quantities in (7.4). Now, we can express (7.2) in the compact form

$$X_i^r = \Sigma_{js} x_{ij}^{rs} + Z_i^r + \bar{E}_i^r \tag{7.6}$$

where the final demand *export* term Z_i^r and external exports $\bar{E}_i^r$ produced at r replaces the satisfied final demand term Y_i^r and exports abroad term E_i^r directly out of r of the classical analysis. This is a *non-trivial* change in the context of the *fixed price* quantity equilibrium implied by classical I-O analysis. In other words, there is no way in (7.5) above to achieve consistency with a final demand consumption vector Y_i^r extracted from an independent demand model. Flow conservation between internal exports and imports to final demand yields the consistency condition

$$\Sigma_r Z_i^r = \Sigma_{rs} y_i^{rs} = \Sigma_{rs} y_i^{sr} = \Sigma_r (Y_i^r - \breve{I}_i^r) \tag{7.7}$$

where $\breve{I}_i^r = \Sigma_s i_i^{sr}$ is the sum of imports of sector i into region r from all regions s. But, this aggregated condition will still not ensure that (7.6) is consistent with an exogenously given final demand Y_i^r and external exports E_i^r. This reasoning shows that there are considerable problems in implementation of interregional I-O models when the usage conditions take into account the outflows to satisfy final demand and external exports in other regions. Of course, in theory, it is finally possible to evaluate the outputs X in a revised matrix form upon substitution of (7.1) into (7.6) via

$$X = (I - A)^{-1} (Z + \bar{E} - \tilde{I}) \tag{7.8}$$

similar to the general *structure* of the conventional model.

We should note in passing that a comparison of the conventional interregional input coefficients a_{ij}^{rs} in region r with a_{ij}^{zs} in another input supply region z yields *no guidance* on the *relative technological effectiveness* of obtaining a unit of *output* of sector j in region s from a unit of *input* of sector i from region r *vs.* a unit of the same *input* from another region z. In order to achieve such a comparison, we would need to distinguish the part of the output X_j^{rs} which is produced from inputs of sector i from region r versus that X_i^{zs} from region z, and have enhanced coefficients $\hat{a}_{ij}^{rs}$ defined as

$$\hat{a}_{ij}^{rs} = x_{ij}^{rs} / X_j^{rs} \tag{7.9}$$

Usually, it is not feasible in practice to obtain output information differentiated by the region of origin of the inputs. Thus, it seems to this writer that the gains from interregional analysis may not be substantial. It appears to yield nothing to enable comparison of the technology embedded in the *inputs* from different regions regarding their *output* efficiency in different sectors in the same or other destination regions, which is what a true *interregional production function* should be trying to measure between regions![2] On the other hand, it does allow us to compare the ratios of local inputs vs. the inputs from other regions, specialised for each different producing sector in a region. Although, from the limitations discussed above, the conventional interregional analysis averages out the 'transmission' of *input technology* over all supplying regions, it can measure the full set of commodity to sector flows between each pair of regions. At the same time, the point remains that most commodity flow data between regions just refers to the commodity being shipped, irrespective of the sectors to which it is to be supplied as intermediate input or final demand. Also, even our enhanced analysis (7.8) is incapable of evaluating the influences on flows of changes in transport costs or networks. Thus, we turn to more useful approaches.

7.1.2 Probabilistic Multi-Regional Flows with Multi-Regional Data

The first step in proceeding from Sect. 7.1.1 is to define input-output relations in terms of total measurable flows. In fact, we define the *total* multi-regional measurable flow $\bar{x}_i^{rs}$ of sector i between regions r and s as

$$\bar{x}_i^{rs} = \Sigma_j (x_{ij}^{rs} + i_{ij}^{rs}) + y_i^{rs} + e_i^{rs} + i_i^{rs} \tag{7.10}$$

which *pool* the intermediate internal and external flows to all sectors j, the internal flows to final demand plus the external export flows and import flows direct to final demand. Note that, this sectoral aggregation over the destination sectors of the flows of a good between each pair of regions, based here entirely on data availability, is a more general form of the *supply pool* assumption of Leontief-Strout. Then, we represent the overall *output technology* by defining multi-regional coefficients a_{ij}^s, denoting the total number of units $x_{ij}^s = \Sigma_r x_{ij}^{rs}$ and $i_{ij}^s = \Sigma_r i_{ij}^{rs}$ of internal and external intermediate inputs of sector i going into sector j in region s from all regions r required to produce a unit of sector j output in the same region s, given as

$$a_{ij}^s = (x_{ij}^s + i_{ij}^s) / X_j^s \tag{7.11}$$

where comparison with (7.1) yields

---

[2] Perhaps these points are obvious to the experienced input-output specialist. However, they do not appear to be greatly emphasised in the literature.

$$a_{ij}^{s} = \Sigma_r a_{ij}^{rs} \tag{7.12}$$

Referring to Leontief and Strout (1963), it is clear that the simplification of the conventional interregional relations (7.1) to the multi-regional relations (7.11) corresponds to their *demand pool* assumption. The next step is to perform some aggregation on the interregional relations. Because the demand-driven relations (7.6) relate to *usage* of the output in terms of *outflows*, we can develop a viable way to eliminate the unknown outputs from the analysis by transposing the basic interregional relations (7.1), reversing the r and s indices and summing over j and s to yield the following *inflow* relations of goods i to all sectors j in region r via

$$\Sigma_{js} (x_{ij}^{sr} + i_{ij}^{sr}) = \Sigma_{js} a_{ij}^{sr} X_j^r \tag{7.13}$$

From (7.12), with the r and s indices reversed, this simplifies to

$$\Sigma_{js} (x_{ij}^{sr} + i_{ij}^{sr}) = \Sigma_j a_{ij}^{r} X_j^r \tag{7.14}$$

The next step is the elimination of the outputs X_j^r from (7.14) in terms of exogenous base period quantities via representation of (7.6) in terms of the measurable flows $\bar{x}_i^{sr}$, as well as eliminating the intermediate inputs on the left-hand side via use of (7.3) and (7.10). From (7.10), $\Sigma_{js} (x_{ij}^{sr} + i_{ij}^{sr}) = \Sigma_s (\bar{x}_i^{sr} - y_i^{sr} - e_i^{sr} - i_i^{sr})$. From (7.3) and (7.5), this simplifies at the base period to $\Sigma_{js} (x_{ij}^{sr} + i_{ij}^{sr}) = \Sigma_s \bar{x}_i^{sr} - Y_i^{r0} - E_i^{r0}$. By substituting this into (7.14), as well as using (7.4), (7.5), (7.6) and (7.10) [reversing the i and j indices], this finally yields (7.14) as

$$\Sigma_s \bar{x}_i^{sr} - \Sigma_j a_{ij}^{r} (\Sigma_s \bar{x}_j^{rs}) = [(Y_i^{r0} + E_i^{r0}) - \Sigma_j a_{ij}^{r} I_j^{r0}] \tag{7.15}$$

If we omit the external exports E_j^{r0} and external imports I_i^{r0}, equation (7.15) turns out to be identical to the LS balance relations. This is surprising as well as comforting, as we have introduced the complete set of flows explicitly into the analysis. At the same time, the weighted imports in the second term on the right-hand side of (7.15) have no counterpart in the LS analysis. In contrast to the awkward situation for the interregional model in (7.8), the true final demand and external export and import quantities have re-emerged in (7.15), allowing the model to be consistent with exogenous final demand, exports and imports.

Armed with the above balance relations (7.15), we can now proceed further with development of a probabilistic model. A considerable enhancement of the Wilson (1970) approach is made by ensuring that the flows reflect the current technology by inserting the enhanced technological balance relations (7.15), setting logistic constraints on regional output capacity and modelling the total measurable flows. Whereas the fundamental form of the conventional entropy model in the absence of economic information would yield equal sectoral production in each region, the enhanced model yields equal *relative capacity utilisation*, a more plausible hypothesis. Also, a strong distinction is made between the estimation form of the model and a transformed version for projection, building on the formalism introduced by Lesse (1982) and specialised to this class of problem by the author (see Sects. 3.2.3 and 3.2.4).

Consider that we have base period data, both on the *total output capacity* $\tilde{X}_{ir}^0$ of sector i to supply intermediate inputs, final demand and external exports from region r, as well as the total in-flow $\overline{X}_s^0$ of all sectors into region s. Also, let $\overline{X}^0$ be the total value of all inputs in the system, $\bar{c}^0$ be the average transport cost per value unit of commodity shipped and c_i^{rs0} be the transport cost per value unit of sector i between regions r and s. Note that, by normalizing with respect to prices, transport costs could be converted into *quantity* units. As the total flows $(\Sigma_s\ \overline{x}_i^{rs})$ in (7.10) include contributions from the imports I_i^r which are not directly related to usage of the output, the model must subtract these from the total in looking at the overall output capacity. Thus, the number of microstates Z can now be given as the number of ways $\tilde{X}_{ir}^0$ distinguishable output capacity units may be divided into $(\Sigma_s \overline{x}_i^{rs} - I_i^{r0})$ which are utilised and $[\tilde{X}_{ir}^0 - (\Sigma_s \overline{x}_i^{rs} - I_i^{r0})]$ which remain unutilised, times the ways that the $(\Sigma_s \overline{x}_i^{rs} - I_i^{r0})$ distinguishable utilised units may be allocated to regions s, yielding

$$Z = \pi_{ir} \{ \tilde{X}_{ir}^0 ! / [\tilde{X}_{ir}^0 - (\Sigma_s \overline{x}_i^{rs} - I_i^{r0})] ! (\Sigma_s \overline{x}_i^{rs} - I_i^{r0})! \} \,.\, \pi_{ir} \{ (\Sigma_s \overline{x}_i^{rs} - I_i^{r0})! / (\pi_s [\overline{x}_i^{rs} - (i_i^{rs0} + \Sigma_j i_{ij}^{rs0})] !) \} \tag{7.16}$$

where, as usual, π denotes the product sign. After some cancellation, setting the entropy S as the natural log of Z and applying the Stirling approximation, we obtain the entropy maximization objective

$$S = - \Sigma_{irs} [\overline{x}_i^{rs} - (i_i^{rs0} + \Sigma_j i_{ij}^{rs0})] \{\log [\overline{x}_i^{rs} - (i_i^{rs0} + \Sigma_j i_{ij}^{rs0})] - 1\} - \Sigma_{ir} [\tilde{X}_{ir}^0 - (\Sigma_s \overline{x}_i^{rs} - I_i^{r0})] \{\log [\tilde{X}_{ir}^0 - (\Sigma_s \overline{x}_i^{rs} - I_i^{r0})] - 1\} \tag{7.17}$$

In relation to the above objective, it is important to remain consistent with total destination in-flow information, yielding the constraints

$$\Sigma_{ir}\ \overline{x}_i^{rs} = \overline{X}_s^0 \tag{7.18}$$

The logistic capacity 'constraint' in the second entropy term in (7.17) allows the usual out-flow constraints to be omitted. Now, apply the average transport cost constraint over the entire system as

$$\Sigma_{irs}\ \overline{x}_i^{rs}\ c_i^{rs0} = \bar{c}^0\ \overline{X}^0 \tag{7.19}$$

Finally, the technological balance relations (7.15) are imposed. With final demand Y_i^{r0}, exports E_i^{r0} and imports I_i^{r0} taken as exogenous inputs, the right-hand side of (7.15) is a known value, as required. Applying Lagrangian theory, (7.17) is maximized under (7.18) with multipliers λ_s, (7.19) with multiplier β and (7.15) with multipliers α_{ir}. Differentiation with respect to $\bar{x}_i^{rs}$ and equating to zero gives

$$\bar{x}_i^{rs} = (i_i^{rs0} + \Sigma_j\, i_{ij}^{rs0}) + [\tilde{X}_{ir}^{0} - (\Sigma_s\, \bar{x}_i^{rs} - I_i^{r0})]\ exp - [\lambda_s + \beta c_i^{rs0} + (\alpha_{is} - \alpha_{ir}\, a_{ii}^{r0})] \tag{7.20}$$

Upon transposing, the explicit logistic result emerges

$$\bar{x}_i^{rs} = (i_i^{rs0} + \Sigma_j\, i_{ij}^{rs0}) + \tilde{X}_{ir}^{0}\ exp - [\lambda_s + \beta c_i^{rs0} + (\alpha_{is} - \alpha_{ir}\, a_{ii}^{r0})] \,/\, \{ 1 + (\Sigma_s\, exp - [\lambda_s + \beta c_i^{rs0} + (\alpha_{is} - \alpha_{ir}\, a_{ii}^{r0})\,]\} \tag{7.21}$$

in which the initial external imports term represents the total external imports of sector i coming into region r which are destined for region s (such 3-index import data would need to be available to implement the model). Setting $A_s = exp - \lambda_s$, $B_{ir} = \tilde{X}_{ir}^{0} / \{1 + (\Sigma_s\, A_s\, D_{is}\, G_{ir}\, exp - \beta c_i^{rs0})\}$, $G_{ir} = D_{ir}^{-a_{ii}^{r0}}$ with $D_{is} = exp - \alpha_{is}$, we can substitute into the constraints. With B_{ir} and G_{ir} already given explicitly above in terms of the other unknowns, we merely need to give expressions for A_s and D_{ir}, which after setting $F_{is} = (\Sigma_r\, B_{ir}\, G_{ir}\, exp - \beta c_i^{rs0})$, come out as

$$A_s = (\,\bar{X}_s^{0} - I^{s0})\, \Sigma_i (D_{is}\, F_{is})\,; \quad D_{ir} = \{\, (Y_i^{r0} + E_i^{r0} - \bar{I}_i^{r0}) + \Sigma_j\, a_{ij}^{r}\, B_{jr}\, G_{jr}\, \Sigma_s\, A_s\, D_{js}\ exp - \beta c_j^{rs0}]\} \,/\, [A_r F_{ir}] \tag{7.22}$$

It is noticed that the expression for D_{ir} is a function of itself, requiring an iteration within an iteration. The gravity parameter β can be determined simultaneously using linear extrapolation. Alternatively, the non-linear equations for both D_{ir} and β could be solved simultaneously using the Newton-Raphson approach, with those for A_s, B_{ir} and E_{ir} being solved using the conventional successive substitution technique. Convergence is guaranteed (if data is consistent) for this problem with a strictly concave objective and linear constraints. The flows are given simply as

$$\bar{x}_i^{rs} = (i_i^{rs0} + \Sigma_j\, i_{ij}^{rs0}) + A_s\, B_{ir}\, D_{is}\, G_{ir}\, exp - \beta c_i^{rs0} \tag{7.23}$$

Although (7.23) may appear to be a *separable* function, connectivity is embedded in the relations (7.22). In fact, the above solution process illustrates the *interdependencies* between transport costs, production technology and logistic capacity constraints in the determination of the flows. After the commodity flows $\bar{x}_i^{rs}$ are obtained from the above, the regional *outputs* X_i^r can be found from the multi-regional form of (7.6), using (7.4) as shown above (7.15). Also, relation (7.11) allows the total intermediate *inter-sectoral usage* x_{ij}^{s} (that is, $\Sigma_r\, x_{ij}^{rs}$) in region s to be evaluated, consistent with the multi-regional technology. **Thus, our**

model yields intermediate inter-sectoral usage within regions and complete flows of a given commodity (sector) between regions. At this estimation step, the modelled flows $\bar{x}_i^{rs}$ should be compared with the observed base period flows $\bar{x}_i^{rs0}$ using a goodness of fit measure, such as Root Mean Square. If the fit is good, the entropy approach will be adequate, especially if the goodness of fit of the model is also accompanied by a significant reduction in the uncertainty of the distribution. Otherwise, one should adopt the information theory procedure described later.

7.1.3 Probabilistic Interregional Flows with Multi-Regional Data

Although this approach may seem like 'getting something for nothing', we use more disaggregated forms of the constraints to formulate an interregional model (with *i, j, r* and *s* flow indices) based on the same more readily available multi-regional flow and technology data used in the previous section. Models in the tradition of Chenery (1953) were able to apply trade coefficients to yield *interregional flows*, despite the limitations of a multi-regional technology. The above approach is modified here to do the same. Also, this turns out to again remove the anomaly in Sect. 7.1.1, where consistency with an exogenous final demand could not be attained. At the same time, we obtain not only the usual intermediate input flows x_{ij}^{rs}, but also the final demand flows y_i^{rs}.

The first step is to disaggregate the balance relations (7.14) in the form

$$\Sigma_s (x_{ij}^{sr} + i_{ij}^{sr}) = (\Sigma_s a_{ij}^{sr0}) X_j^r = a_{ij}^r X_j^r \tag{7.24}$$

Then, using (7.2) with the reversal of the i and j indices, we can eliminate X_j^r from (7.23) and transpose to yield the balance relations, the counterpart to (7.15), as

$$\Sigma_s x_{ij}^{sr} - a_{ij}^r (\Sigma_{is} x_{ji}^{rs} + \Sigma_s y_j^{rs}) = a_{ij}^r \bar{E}_j^{r0} - \bar{I}_{ij}^{r0} \tag{7.25}$$

where the newly defined quantity $\bar{I}_{ij}^{r0}$ represents the imports of sector i into region r from all regions s as intermediate inputs into sector j.

In formulating the entropy, we realise that the output capacity $\tilde{X}_{ir}^0$ for sector i in region r is available for producing both intermediate inputs, direct inputs to final demand and exports. Then, the unutilised capacity is here expressed as $(\tilde{X}_{ir}^0 - \bar{E}_i^{r0} - \Sigma_{js} x_{ij}^{rs} - \Sigma_s y_i^{rs})$, as seen from (7.2). The entropy S in (7.17) now becomes S' as

$$S' = - \Sigma_{ijrs} x_{ij}^{rs} (\log x_{ij}^{rs} - 1) - \Sigma_{irs} y_i^{rs} (\log y_i^{rs} - 1) - \Sigma_{ir} (\tilde{X}_{ir}^0 - \bar{E}_i^{r0} - \Sigma_{js} x_{ij}^{rs} - \Sigma_s y_i^{rs}) \\ [\log (\tilde{X}_{ir}^0 - \bar{E}_i^{r0} - \Sigma_{js} x_{ij}^{rs} - \Sigma_s y_i^{rs}) - 1] \tag{7.26}$$

Constraints (7.18) are disaggregated to yield

$$\Sigma_{jr} x_{ij}^{rs} = \overline{X}_{is}^{0} - E_i^{s0} - Y_i^{s0} - \tilde{I}_i^{s0} \tag{7.27}$$

where $\overline{X}_{is}^{0} = \Sigma_r \overline{x}_i^{rs}$ represents the measurable inflow into region s of all goods i, for intermediate inputs, final demand, exports and imports.[3] The transport constraint (7.19) now becomes

$$\Sigma_{ijrs} x_{ijrs} c_i^{rs0} + \Sigma_{irs} y_i^{rs} c_i^{rs0} = c^0 \overline{X}^0 - C_I^0 - C_E^0 \tag{7.28}$$

where C_I^0 and C_E^0 are the respective base period total internal transport costs of external imports and exports - this data may not be easy to acquire. Finally, two extra sets of constraints are added. The first is on final demand net of external imports via (7.3) and (7.7)

$$\Sigma_s y_i^{sr} = Y_i^{r0} - \tilde{I}_i^{r0} \tag{7.29}$$

As we may not know the base period internal final demand exports Z_i^{r0} [see (7.4)], we rely on the internal import/export conservation conditions $\Sigma_{rs} y_i^{sr} = \Sigma_{rs} y_i^{rs} = Y_i^0 - \tilde{I}_i^{r0}$ to write analogous constraints via (7.7)

$$\Sigma_{rs} y_i^{rs} = Y_i^0 - \tilde{I}_i^0 \tag{7.30}$$

With both the objective function and the constraints now completely defined, we can perform our maximization as before, yielding

$$x_{ij}^{rs} = (\tilde{X}_{ir}^{0} - \overline{E}_i^{r0} - \Sigma_{js} x_{ij}^{rs} - \Sigma_s y_i^{rs}) \exp - (\lambda_{is} + \beta c_i^{rs0} + \alpha_{ijs} - \alpha_{jir} a_{ji}^r) \tag{7.31}$$

where λ_{is} is the Lagrange multiplier on constraint (7.27), β that on (7.28) and α_{ijr} that on (7.25). Similarly, the flows y_i^{rs} to final demand emerge as

$$y_i^{rs} = (\tilde{X}_{ir}^{0} - \overline{E}_i^{r0} - \Sigma_{js} x_{ij}^{rs} - \Sigma_s y_i^{rs}) \exp - (\beta c_i^{rs0} + \Sigma_j \alpha_{jir} a_{ji}^r + \gamma_{is} + \eta_i) \tag{7.32}$$

where γ_{ir} is the Lagrange multiplier on (7.29) and η_i that on (7.30). As before, we need to define some intermediate terms, with $A_{is} = \exp - \lambda_{is}$, $D_{ijs} = \exp - \alpha_{ijs}$ and $G_{ijs} = D_{ijs}^{a_{ij}^r}$, we can express the second term P_{ijrs} of (7.31) as $P_{ijrs} = A_{is} D_{ijs} G_{jir} \exp - \beta c_i^{rs0}$. Similarly, setting $H_{is} = - \gamma_{is}$ and $L_i = \exp - \eta_i$, we write the second term of (7.31) as $Q_{irs} = H_{is} L_i (\pi_j G_{jir}) \exp - \beta c_i^{rs0}$. We now eliminate the x and y

---

[3] Another option was to replace these constraints by new output balance relations based on the row sum purchases plus value added. However, to be viable, this needed to also include supply-driven coefficients, making the resulting model much less transparent.

terms from the right-hand sides of (7.31) and (7.32), upon which we define $B_{ir} = (\tilde{X}_{ir}^{0} - \bar{E}_i^{r0}) / [1 + \sum_{js} P_{ijrs} + \sum_s Q_{irs}]$. This allows us to finally write

$$
\begin{aligned}
x_{ij}^{rs} &= B_{ir} A_{is} D_{ijs} G_{jir} \exp - \beta c_i^{rs0} \\
y_i^{rs} &= B_{ir} H_{is} L_i (\pi_j G_{jir}) \exp - \beta c_i^{rs0}
\end{aligned}
\tag{7.33}
$$

where π_j refers to the product over j. Upon substitution into the constraints, we get

$$
\begin{aligned}
A_{is} &= (\bar{X}_{is}^{0} - E_i^{s0} - Y_i^{s0} - \tilde{I}_i^{s0}) / [\textstyle\sum_{jr} B_{ir} D_{ijs} \exp - \beta c_i^{rs0}] \;\; ; \;\; H_{ir} = (Y_i^{r0} - \breve{I}_i^{r0}) \\
&/[\textstyle\sum_s B_{is} L_i (\pi_j G_{jis}) \exp - \beta c_i^{rs0}]; \; L_i = (Y_i^{0} - \breve{I}_i^{0}) / [\textstyle\sum_{rs} B_{ir} H_{is} (\pi_j G_{jir}) \exp - \beta c_i^{rs0}] \;\; ; \\
D_{ijr} &= \{ a_{ij}^{r} (E_j^{r0} + \textstyle\sum_{is} B_{jr} A_{js} D_{jis} G_{ijr} \exp - \beta c_j^{rs0} + \sum_s B_{jr} H_{js} L_j (\pi_i G_{ijr}) \exp \\
&\quad - \beta c_j^{rs0}) - \bar{I}_{ij}^{r0} \} / [\textstyle\sum_s B_{is} A_{ir} G_{jis} \exp - \beta c_i^{rs0}]
\end{aligned}
\tag{7.34}
$$

The iterative solution procedure is similar to that detailed below (7.21) and (7.22).

Although no further technological and transport information is provided over (7.23), we can now obtain information on the regional mix of inputs and flows directly to final demand, with just measurable flow quantities required as estimation input. In principle, such detailed results represent a bonus.

7.1.4 Implementation Issues

Use of models for projection. The main task in transforming the estimation model formulations in Sects. 7.1.2 and 7.1.3 into a form suitable for projection is to choose the Lagrange multipliers which should be treated as parameters in projection, and those which must be evaluated anew. This clearly relates to which information should reasonably be treated as exogenous input and which treated as endogenous output. In terms of the structure of the model, it is considered that the exogenous input should include any quantity changes, such as changed regional output capacities $\tilde{X}_{ir}$, changed external exports E_i^r, changed external imports I_i^r and changed final demand Y_i^r. If transport costs change to c_i^{rs}, the availability as a parameter of the gravity Lagrange multiplier β allows the influence of these changes to be assessed by the model. The formalism for this was introduced by (Lesse, 1982) and first applied generally to spatial interaction models by the author, as described in Sects. 3.2.3 and 3.2.4. In addition, any new multi-regional input-output coefficients a_{ij}^{r} must be provided exogenously. Finally, the treatment of the Lagrange multipliers λ_s or λ_{is} as a parameter allows the in-flows to be endogenous in projection, in response to the changes in output capacities, final demand, transport costs and the multi-regional technology. In this case, the projection form of (7.21) changes to

$$
\begin{aligned}
\bar{x}_i^{rs} &= (i_i^{rs} + \textstyle\sum_j i_{ij}^{rs}) + \tilde{X}_{ir} \exp - [\lambda_s + \beta c_i^{rs} + (\alpha_{is}' - \alpha_{ir}' a_{ii}^{r})] / \{ 1 + \\
&\quad (\textstyle\sum_s \exp - [\lambda_s + \beta c_i^{rs} + (\alpha_{is}' - \alpha_{ir}' a_{ii}^{r})] \}
\end{aligned}
\tag{7.35}
$$

with merely the multipliers α_{is}' having to be re-evaluated to satisfy an updated form of the balance relations (7.15). The evaluation of these multipliers would just entail successive substitution into the relations associated with (7.21), but where A_s and β are now given exogenously. Similar reasoning applies for the interregional model (7.31) and (7.32), where in this case we re-evaluate the multipliers on the more disaggregated balance relations (7.25) to α_{ijr}'. This recursive substitution process retains the non-separable structure of the models. Furthermore, it is clear that our probabilistic approach has bypassed the necessity of matrix inversion, which is routinely required for the deterministic procedure.

An adjustment from information theory. As stressed by Batten (1983), the more general information theory approach may give improved prediction ability for flow models of this type. If the goodness of fit of the estimated models is not satisfactory, the projection model is likely to yield improved results if information bias terms are computed and inserted into the models, as demonstrated via a new information theory procedure in Roy (1987). We proceed as follows. If, for example, the estimated entropy model (7.21) yields x_i^{rs*} as the modelled flows, we compute ratios z_i^{rs} as the quotient $(\bar{x}_i^{rs0} / x_i^{rs*})$ of the observed and modelled flows and z_i^r as the quotient $[\tilde{X}_{ir}^0 - (\Sigma_s \bar{x}_i^{rs0} - I_i^{r0})] / [\tilde{X}_{ir}^0 - (\Sigma_s x_i^{rs*} - I_i^{r0})]$ of the observed and modelled unutilised capacities. The next step is to normalise the z's as probabilities $q_i^{rs} = z_i^{rs} / (\Sigma_{irs} z_i^{rs})$ and $q_i^r = z_i^r / (\Sigma_{ir} z_i^r)$. If the q's are applied as prior probabilities in the denominator of the log terms in the entropy objective (7.16) and this revised estimation problem is solved anew, it is demonstrated in Roy (1987) that (i) all the Lagrange multipliers are (reasonably) *invariant* to this changed objective and (ii) the modelled flows *precisely* correspond to the observed flows $\bar{x}_i^{rs0}$. Of course, this is a good basis for projection. In addition, the outflow and gravity parameters are retained, with the latter allowing the model to be sensitive to future transport cost and network changes. The consistency of the flows with the multi-regional technology available at the projection period is assured by applying the technological balance relations anew. The enhanced projection relations, as a modification of (7.35), now become

$$\bar{x}_i^{rs} = (i_i^{rs} + \Sigma_j i_{ij}^{rs}) + \tilde{X}_{ir} \, q_i^{rs} \, exp - [\lambda_s + \beta c_i^{rs} + (\alpha_{is}'' - \alpha_{ir}'' a_{ii}^r)] / \{ q_i^r + (\Sigma_s q_i^{rs} \, exp - [\lambda_s + \beta c_i^{rs} + (\alpha_{is}'' - \alpha_{ir}'' a_{ii}^r))]\} \quad (7.36)$$

where the Lagrange multipliers α_{ir}'' associated with the new input-output balance constraints will change from those in the pure entropy formulation (7.35). It is suggested that (7.36) be used in place of (7.35) for most applications. Similar reasoning applies for the interregional model in Sect. 7.1.3.

Some Comparisons. The approaches presented in Sects. 7.1.2 and 7.1.3 integrate technology changes, input capacity changes and transport network and cost changes in the evaluation of multi-regional flows. As probabilistic models, they fit parameters to observations, rather than relying on deterministic optimization. The logistic form of the supply relationships promotes spillovers into adjacent regions when there is both a vintage distribution of capacity a high pressure on this capacity in a given region. At the same time, the models do not attempt to project changes in technology. It is the user's responsibility to provide any changed I-O coefficients a_{ij}^{r}. Their primary aim is to estimate the influence on regional flows and output when exogenous changes in input capacities, final demand, transport costs and technical coefficients are introduced. They will both need to be tested with conventional multi-regional data to verify their suitability for these tasks. These data will need to be enhanced by capacity data for each sector in each region. Without the technological balance relations, the model would resemble a commodity flow model enhanced by the logistic input capacity constraints. However, we maintain that the flows must be consistent with the multi-regional technology being used, justifying the enforcing of these balance relations. The Leontief-Strout balance relations, which were also used by Wilson (1970), are shown to be correct if enhanced by terms related to external imports and exports.

What if we do not have sector output capacity data? Data on the net sector output capacities $\tilde{X}_{ir}^{0}$ for each sector in each region may be quite difficult to acquire. Also, the concept of capacity may be quite 'fuzzy' in situations of high demand, where some dormant or outmoded plants may well be called into emergency production. In that case, base period constraints are introduced on the total value $\overline{X}_{r}^{0}$ of out-flows from each origin region r, in the form

$$\Sigma_{is}\ \overline{x}_{i}^{rs} = \overline{X}_{r}^{0} \qquad (7.37)$$

where $\overline{X}_{r}^{0}$ equals the sum $(X^{r0} + I^{r0})$ of the total value of production at region r minus the total imports directly into region r. If the capacity entropy is omitted as the second term in (7.17), and (7.37) attached with a multiplier η_r, then the previous solution (7.21) is replaced by

$$\overline{x}_{i}^{rs} = exp - [\lambda_s + \eta_r + \beta c_i^{rs0} + (\alpha_{is} - \alpha_{ir}\, a_{ii}^{r0})] \qquad (7.38)$$

With A_s, D_{is} and G_{ir} defined in (7.22) and B_r defined as $exp - \eta_r$, we can set up the following set of recursive relations

$$A_s = (\overline{X}_{s}^{0} - I^{s0}) / \Sigma_{ir} B_r D_{is} G_{ir}\, exp - \beta c_i^{rs0}\ ; \quad B_r = \overline{X}_{r}^{0} / \Sigma_{is} A_s D_{is} G_{ir}\, exp - \beta c_i^{rs0}$$
$$D_{ir} = \{ (Y_i^{r0} + E_i^{r0} - \overline{I}_i^{r0}) + \Sigma_j a_{ij}^{r} [E_j^{r} + B_r G_{jr} \Sigma_s A_s D_{js}\ exp - \beta c_j^{rs0}]\} / \{\Sigma_s A_r B_s$$
$$G_{is}\ exp - \beta c_i^{sr0}\} \qquad (7.39)$$

These three relations, together with that for G_{ir}, can be solved iteratively for the unknown Lagrange multipliers, with β being obtained, as before, via linear extrapolation. The final relations are identical to (7.23) except for B_r replacing B_{ir}

and the additive import correction terms being absent. For projection, A_s, B_r and β are treated as parameters, with the recursive relations for D_{is} and G_{ir} providing the interdependencies associated with the inverse in the conventional solution. Clearly, the information theory adjustment should again be used when the fit to the base period flows is not adequate. Also, as inferred from Johansson (1991), the above alternative form is relevant for regional sectors without a strong vintage distribution over the available production capacity.

7.1.5 Some Further Developments

In Roy (1997b), a probabilistic approach was developed which ***retained prices*** in input-output analysis and allowed the ***supply to final demand*** to be ***endogenous***. Although regional prices are not easily obtained, the provision of prices thus opens the opportunity for input-output models to be developed as short run supply functions, allowing tâtonnement with a compatible demand model in a regional CGE framework. The first change to the interregional model in Sect. 7.1.3 is that the flows x_{ij}^{rs} are defined as in terms of *quantities* shipped, rather than in terms of dollars. The objective function (7.26) remains unchanged. With the supply to final demand being endogenous, the balance relations (7.25) are superfluous here. The transport cost constraint (7.28) should now be modified to a constraint on short run profits, reflecting the classical objective of the competitive firm. If $\overline{\pi}^0$ is taken as the base period unit profit per dollar of all commodities produced, then we should replace (7.28) by a profit constraint in terms of revenue minus costs, yielding

$$\Sigma_{ijrs}(x_{ij}^{rs}+i_{ij}^{rs})\, p_i^{r0} / (m\, a_{ij}^{s0}) \;-\; \Sigma_{irs}(p_i^{r0} + c_i^{rs0})\, [\Sigma_j(x_{ij}^{rs}) + y_i^{rs}] \;=\; \overline{\pi}^0 V^0 \qquad (7.40)$$

where V^0 are the total dollars of production, m the number of row sectors and the p's the fob prices in the base period. Also, assuming that the input-output coefficients a_{ij}^{s0} are given (as usual) in value terms, the output prices p_j^{s0} cancel out in the above revenue expression. In this term, the output has been expressed as the *average* over all rows of the quotients of input divided by the regional I-O coefficient, which emerge as constant values in the deterministic theory (Roy, 1997b). If the Lagrange multiplier on the profit constraint (7.40) is taken as ϕ, then constrained maximization of (7.26) yields for the intermediate inputs

$$x_{ij}^{rs} \;=\; (\,\tilde{X}_{ir}^{0} - \overline{E}_i^{r0} - \Sigma_{js}\, x_{ij}^{rs} - \Sigma_s\, y_i^{rs})\; exp - [\, \lambda_{is} - \phi\,\{(p_i^{r0}/ma_{ij}^{s0}) - p_i^{r0} - c_i^{rs0}\}\,] \qquad (7.41)$$

and for the internal flows to final demand

$$y_i^{rs} \;=\; (\,\tilde{X}_{ir}^{0} - \overline{E}_i^{r0} - \Sigma_{js}\, x_{ij}^{rs} - \Sigma_s\, y_i^{rs})\; exp - [\, \gamma_{is} + \eta_i - \phi\,\{(p_i^{r0}/ma_{ij}^{s0}) - p_i^{r0} - c_i^{rs0}\}\,] \qquad (7.42)$$

If the unknown terms from the left-hand sides are removed, as demonstrated for (7.33), the result truly represents a logistic probabilistic supply function, where the profits are included as a constraint based on observations, rather than as the objective in the deterministic theory. On the other hand, in the projection period, the profit levels become an endogenous output, as should be expected. Output can be obtained as an average of the quotients of intermediate inputs and regional coefficients, consistent with their role in the revenue relations. The supply to final demand can then be computed endogenously from the basic usage relations (7.2). Thus, if regional market prices are available, the relations (7.41) and (7.42) should be used in preference to (7.31) and (7.32), where final demand is exogenous and prices have been 'absorbed', not being capable of modification between the base period and the projection period.

The overall aim of this part of the chapter has been to extend the conventional deterministic theories to be probabilistic and to include all the contributions to the measurable flows. In addition, we can identify the clear interdependencies between output capacities, production technology and transport costs in their influence on the pattern of multi-regional or interregional flows. A further point is the strong emphasis on data availability and its consequences for model structure and the appropriate level of detail in the results. When regional prices are included, regional supply functions are produced, allowing the supply to final demand to be endogenous. At the same time, an apparent anomaly in the classical interregional model has been uncovered. Much empirical work remains to be done to demonstrate the relevance of the proposed alternative model structures. A further challenge, if data is available, is to include the above ideas within a commodity by sector framework. This is especially important for the above models, where transport costs are included explicitly. Within a sector producing several commodities, the unit transport costs for each commodity can vary quite markedly.

7.2 Changes in Both Transport Networks and Production Technology

7.2.1 Motivation for Model

The more general situation in practice is that both the production technology and the transport network (infrastructure or pricing) will change over a certain period. As a consequence, a successful model needs to be able to find a way to represent the base period technology in the model estimation, as well as to allow the changed technology in the current period to be inferred via relevant available economic data, such as regional outputs and final demand. In other words, starting with the base period input-output coefficients and associated flows, we aim to reproduce the revised input-output coefficients in a period where we have observations or independent projections of outputs and final demand in each region, as well as information on the projected state of the transport network. With knowledge via the model results of the set of interregional flows, together with the available regional output, the revised input-output coefficients can be directly evaluated. If we happen to have reliable independent projections of *expected* gross output and final demand by sector and region, the model may then be used to evaluate the expected revised input-output coefficients.

7.2.2 Proposed Interregional Model

Model Estimation. In this model, where the inputs are regional outputs and final demand and the outputs the full set of interregional flows, the use of regional sectoral output *capacities*, as illustrated in the previous section, are superfluous. We merely need to define the classical flow entropy

$$S = -\Sigma_{ijrs} x_{ij}^{rs} \log (x_{ij}^{rs} - 1) - \Sigma_{irs} y_i^{rs} \log (y_i^{rs} - 1) \tag{7.43}$$

The next step is to see how the base period technology can be represented by the model. With knowledge of the base period input-output coefficients a_{ij}^{s0} at the destination regions, as well as the corresponding outputs X_j^{s0} and the intermediate external import flows i_{ij}^{s0}, we can evaluate the implied intermediate internal flows x_{ij}^{s0} via the fundamental relations

$$x_{ij}^{s0} = a_{ij}^{s0} X_j^{s0} - i_{ij}^{s0} \tag{7.44}$$

Thus, in order for the base period model to closely embrace the base period technology, the following constraints are imposed on the interregional in-flows

$$\Sigma_r x_{ij}^{rs} = x_{ij}^{s0} \tag{7.45}$$

In order to include the influence of out-flow information in the model, we transpose the fundamental output usage relations (7.2) to yield

$$\Sigma_{js} x_{ij}^{rs} + \Sigma_s y_i^{rs} = X_i^{r0} - \bar{E}_i^{r0} \tag{7.46}$$

In addition, the final demand inflow and outflow relations, (7.29) and (7.30) respectively, must be imposed. As before, the effects of transport costs are included by application of constraint (7.28).

Then, if the classical flow entropy objective (7.43) is maximized simultaneously in terms of the interregional intermediate flows x_{ij}^{rs} and final demand flows y_i^{rs} under constraints (7.45) with Lagrange multipliers α_{ijs}, constraints (7.46) with multipliers λ_{ir}', constraints (7.29) with multipliers γ_{ir}', constraints (7.30) with multipliers η_i' and constraint (7.28) with multiplier β', the result for the intermediate inputs emerges in the elegant form

$$x_{ij}^{rs} = exp - (\alpha_{ijs} + \lambda_{ir}' + \beta' c_i^{rs0}) \tag{7.47}$$

and for the flows to final demand

$$y_i^{rs} = exp - (\lambda_{ir}' + \gamma_{is}' + \eta_i' + \beta' c_i^{rs0}) \tag{7.48}$$

Note that, both the usage constraints (7.45) and the transport cost constraints (7.28) *couple* the determination of the intermediate input flows x_{ij}^{rs} and the direct final demand flows y_i^{rs} in the analysis. In practice, with the total of output minus final demand being required as input in the projection time period, the Lagrange multipliers λ_{ir}' on (7.46) should be eliminated from (7.47) and (7.48), leading to the result

$$x_{ij}^{rs} = (X_i^{r0} - \bar{E}_i^{r0})\, exp - (\alpha_{ijs} + \beta' c_i^{rs0}) / [\, \Sigma_{js}\, exp - (\alpha_{ijs} + \beta' c_i^{rs0}) + \Sigma_s\, exp - (\gamma_{is}' + \eta_i' + \beta' c_i^{rs0})\,] \tag{7.49}$$

and for the final demand flows

$$y_i^{rs} = (X_i^{r0} - \bar{E}_i^{r0})\, exp - (\gamma_{is}' + \eta_i' + \beta' c_i^{rs0}) / [\, \Sigma_{js}\, exp - (\alpha_{ijs} + \beta' c_i^{rs0}) + \Sigma_s\, exp - (\gamma_{is}' + \eta_i' + \beta' c_i^{rs0})\,] \tag{7.50}$$

The Lagrange multipliers α_{ijs}, γ_{is}', η_i' and β' should be obtained via the solution of the non-linear equations (7.45), (7.29), (7.30) and (7.28) respectively [after substitution of (7.49) and (7.50)] by an iterative approach, such as Newton-Raphson iteration. This procedure is explained in detail within Chap. 1.

Use of Model for Projection. As described previously, our objective is to obtain the revised input-output coefficients when given the current values of output, final demand and transport costs. The formalism of Sect. 3.2.4 can be applied to produce the form of (7.49) and (7.50) for projection. Constraints (7.29), (7.30) and (7.46), representing exogenous values, are re-imposed, with the Lagrange multipliers β' on (7.28) and α_{ijs} on (7.46), where the right-hand sides are needed as exogenous outputs, being treated as parameters. This finally yields for the intermediate inputs

$$x_{ij}^{rs} = (X_i^{r} - \bar{E}_i^{r})\, exp - (\alpha_{ijs} + \beta' c_i^{rs0}) / [\, \Sigma_{js}\, exp - (\alpha_{ijs} + \beta' c_i^{rs}) + \Sigma_s\, exp - (\gamma_{is}'' + \eta_i'' + \beta' c_i^{rs})\,] \tag{7.51}$$

and for the final demand outflows

$$y_i^{rs} = (X_i^{r} - \bar{E}_i^{r})\, exp - (\gamma_{is}'' + \eta_i'' + \beta' c_i^{rs}) / [\, \Sigma_{js}\, exp - (\alpha_{ijs} + \beta' c_i^{rs0}) + \Sigma_s\, exp - (\gamma_{is}'' + \eta_i'' + \beta' c_i^{rs})\,] \tag{7.52}$$

in which the current values of the required observed variables are denoted as X_i^r, E_i^r, Y_i^r and c_i^{rs}, and the Lagrange multipliers γ_{ir}'' and η_i'' are modified from their original values when the corresponding constraints are re-imposed with values different from those at the base period. Having determined the individual interregional intermediate input flows from (7.51), the final step is to evaluate the new multiregional input-output coefficients via the classical relation

$$a_{ij}^s = [\Sigma_r (x_{ij}^{rs} + i_{ij}^{rs})] / X_j^s \tag{7.53}$$

Note that, a more disaggregate form of (7.53) would enable us to obtain the full set of interregional coefficients a_{ij}^{rs} rather than the multi-regional coefficients a_{ij}^s. However, we would really be kidding ourselves if we believed that mere provision of the base period *multi-regional* coefficients a_{ij}^{s0} as data during the model estimation would somehow magically enable us to make reliable projections of the new *interregional* coefficients.

Finally, it is useful to compare the above method with the popular RAS method for updating input-output tables, as described in Sect. 2.2.1. The first difference, when comparing (2.37) with (7.49) to (7.52), is that we use a two-step entropy procedure (estimation model, projection model), rather than the one-step information theory approach implied by RAS. In this regard, we note that it may be very difficult to obtain the base period interregional flow data required for the RAS prior probabilities - we may well have to be content with multi-regional flows. Our different analysis is partly due to our desire to explicitly include the influence of transport costs and changes in the transport infrastructure on the pattern of commodity flows, as well as simultaneously distinguishing the effects of changed technology. The second difference is that, even allowing for expanding RAS into interregional analysis, our outflow constraints (7.45) are defined at a finer level of disaggregation than the corresponding constraints in (2.38), reflecting the full multi-regional information obtainable from the base period technology. Of course, it would be useful in practice to compare the empirical performance of the two approaches.

8 Illustration of Imperfect Competition

8.1 Introduction

In Sect. 6.4, imperfect competition has already been considered in the land/housing market, in a hierarchical scheme, involving landowners, land developers and housing developers. Although such markets, as well as those in the retailing sector, represent situations where imperfect competition is prevalent, many production sectors also exhibit imperfect competition. This chapter addresses imperfect competition in goods production sectors. The first step is the development of demand and supply functions for perfectly competitive markets. Upon this basis, different scenarios for introducing imperfect competition are then introduced, distinguishing between oligopolistic and monopolistic competition as already defined and discussed in Chap. 2. All the developments are classified under the general heading of *dispersed spatial price equilibrium*, a probabilistic enhancement of *spatial price equilibrium*, described in Sect. 3.1.2.

To briefly recapitulate, the classical Spatial Price Equilibrium (SPE) models of Takayama and Judge (1971), described in Sect. 2.2.2, possesses an exchange (distribution) mechanism allocating flows to a given destination only from the origin with the minimum cif (delivered) price. Whist such a deterministic structure is useful for certain bulk commodities, such as agricultural goods, it is much less appropriate for more differentiated goods (for example, automobiles). As defined by Batten and Westin (1989), the *dispersed* exchange mechanisms allow spillovers onto alternative links, as well as permitting cross-hauling (that is, goods being exported from region i to region j, in addition to being exported from region j to region i). Rather than incorporating the optimization properties and homogeneity assumptions of the classical SPE models, the *dispersed* (DSPE) versions postulate certain *distributions* of exchange behaviour about an endogenous average for each region, with their parameters obtainable in terms of best-fit criteria to observed patterns of flows, costs and prices. Alternatively, identical model forms are derivable via entropy maximization techniques. Both approaches are presented in the following.

The DSPE models described in Sect. 3.1.2 concentrate on developing an exchange mechanism to link exogenously given supply and demand functions in the different regions. These were usually described as separable. In Roy and Johansson (1993), spatial supply and demand models were developed endogenously for all agents, with importers introduced as explicit profit-maximising intermediaries, replacing the abstract exchange mechanisms of previous models. Certain distributions across the supply and demand capacities of each region were introduced to maintain non-separability, as illustrated in principle in Sect. 3.2.6. Differentiated goods were handled by identification of the willingness-to-pay for characteristics of the goods.

In this chapter, we account simultaneously for both differentiated goods and differentiated markets in a direct bargaining framework which omits the exchange mechanism. Instead, spatial linkages are incorporated directly into the unified formulation of the non-separable spatial supply and demand models, resulting in link-specific market prices and corresponding link-specific equilibria. At that point, imperfect competition is introduced within the DSPE framework, building on a similar generalisation of the classical SPE framework achieved by Kolstad and Mathieson (1991) and Nagurney and Dupuis (1993).

8.2 A General Equilibrium Framework

8.2.1 A Model of Producer Behaviour

Distributional Approach. In Roy and Johansson (1993), a separable producer model was introduced, with fob prices (that is, buyers pay transport costs) p_r in each producing region r. In practice, the fob prices p_r represent averages of link-specific fob prices p_{rs} negotiated in a *mill pricing* format with the different demand regions s. If this price variance over s cannot be neglected, it is necessary to replace the simpler producer model in Roy and Johansson with a model with link-specific prices p_{rs} which is also non-separable. This price variation can reflect the relative bargaining power of the alternative demand regions in relation to factors such as their size, their different living standards and different levels of import tariffs or transport costs. This more general model is now developed.

Consider a system with m demand regions (countries) and n supply regions (countries). With the production level in any region r denoted as X_r, one may not, in practice, have the individual establishment and input requirement data to develop regional marginal cost functions $v_r\ (X_r)$. Instead, one may just have estimates of the *average* costs $\bar{v}_r$. However, in cases where production costs and product quality are strongly segmented in relation to supplying the alternative destination regions s, one may be able to identify link-specific average production costs $\bar{v}_{rs}$ yielding short run unit profits π_{rs} as

$$\pi_{rs} = p_{rs} - \bar{v}_{rs} \tag{8.1}$$

In deciding from each region r how the ratio of the actual flow x_{rs} supplied to any region s over the expected flow (X_r/m) behaves, we postulate the following plausible log-linear relationship in relation to average expected profits, where we express the log of quantity ratios in terms of corresponding profit deviations

$$log\,(m\, x_{rs}/X_r) \;=\; \nu(\pi_{rs} - \tilde{\pi}_r - \Delta\tilde{\pi}_s) \tag{8.2}$$

where $\tilde{\pi}_r$ defines a perceived expected average profit for active producers at r, $\Delta\tilde{\pi}_s$ is a profit 'correction' parameter with respect to the satisfaction of observed demands from all regions s and ν is an 'elasticity' parameter. Although this relationship would be quite sufficient to estimate (for example, via maximum likelihood) a *supply share* model, it possesses no device for endogenous determination of the regional production levels X_r.

As discussed in Sect. 3.2.6, Hotelling (1932) presented cogent arguments for the existence of logistic forms of regional supply and demand functions. These arguments were related to the form of *vintage* supply functions by Johansson (1991). Faced with the prospect of the availability of merely average regional cost data and desiring a truly price-elastic model with the regional production X_r determined endogenously, a logistic supply function may be obtained by postulating a relationship consistent with (8.2) across the shares of utilised capacity X_r vs. unutliised capacity $(\overline{X}_r - X_r)$ for each region r, where $\overline{X}_r$ is the short run *available production capacity* of the particular commodity sector in r. If a perceived average profit $\overline{\pi}_r$ is defined across the total available capacity, $\tilde{\pi}_r$ is taken to apply to the active capacity (see above) and zero profit to the unutilised capacity (neglecting holding costs), one may again postulate log-linear actual to expected ratios, in analogy to (8.2) in the form

$$\begin{aligned} log\,[X_r/\{\overline{X}_r/2\}] &= \nu(\tilde{\pi}_r - \overline{\pi}_r) \\ log\,[(\overline{X}_r - X_r)/\{\overline{X}_r/2\}] &= \nu(0 - \overline{\pi}_r) \end{aligned} \tag{8.3}$$

which, upon elimination of $\overline{\pi}_r$, yields

$$log\,[X_r/(\overline{X}_r - X_r)] \;=\; \nu\,\tilde{\pi}_r \tag{8.4}$$

Together with (8.2), this results in a logistic price-responsive model as justified by Hotelling. The ratio $\overline{X}_r/(\overline{X}_r - X_r)$ is commonly known as *capacity tension*, and $X_r/(\overline{X}_r - X_r)$ as the *odds ratio*.

Note that, the same sensitivity parameter ν has been used for both of the share models (8.2) and (8.3). However, it can be shown (Roy, 1985) that, whereas this simplification yields a model structure similar to *multinomial logit*, the use of different parameters in (8.2) and (8.3) would have implied a *nested logit* structure.

If the consistency parameter $\tilde{\pi}_r$ is eliminated by substitution of the logistic relation (8.4) into the supply share relations (8.2), the complete model emerges as

$$x_{rs} \;=\; \overline{X}_r\;(1/m)\;exp\;\nu(\pi_{rs} - \Delta\tilde{\pi}_s)\,/\,[1 + \Sigma_s\,(1/m)\;exp\;\nu(\pi_{rs} - \Delta\tilde{\pi}_s)] \tag{8.5}$$

If the system (8.2) and (8.4) is estimated via the maximum likelihood method, it emerges that $\tilde{\pi}_r$ can be estimated to satisfy the supply consistency conditions

$$\Sigma_s x_{rs} = X_r \tag{8.6}$$

The parameter $\Delta\tilde{\pi}_s$ ensures that the observed base period demand D_s^0 in each region s is reproduced by the model via

$$\Sigma_r x_{rs} = D_s^0 \tag{8.7}$$

The parameter ν guarantees that the observed base period profits $(\Sigma_{rs} x_{rs}^0 \pi_{rs})$ are reproduced by the model in the form

$$\Sigma_{rs} x_{rs} \pi_{rs} = \Sigma_{rs} x_{rs}^0 \pi_{rs} \tag{8.8}$$

In Appendix 1.1, we illustrated the maximum likelihood evaluation of the floorspace index α in the classical retail model. As the parameter evaluation for the more sophisticated model above is rather more complicated, it is demonstrated in Appendix 8.1 at the end of this chapter.

Even with link-specific prices p_{rs}, the supply function (8.6) maintains *non-separability*, that is, $\partial x_{rs} / \partial p_{rv} \neq 0$ when $r \neq v$. Also, it is possible to optionally replace the uniform 'elasticity' ν by regional-specific values ν_r, whereby the estimated model would reproduce the base period profits $(\Sigma_s x_{rs}^0 \pi_{rs})$ in each producing region r. As seen later, such constraints are consistent with treatment of each region r as an oligopolist.

Entropy Approach. As the primary emphasis in this book is the formulation of entropy maximisation models, we illustrate an entropy derivation of the supply model illustrated in the previous section.

Let the flows x_{rs} represent individual (distinguishable) contracts between producers in regions r and s, emanating from the total available productive capacity $\overline{X}_r$ of region r and flowing to region s as a component of the observed base period demand D_s^0, expressed here via (8.7). These are interpreted here as constraints, rather than as the maximum likelihood equations associated with a specified parameter. We may bypass the parameters $\overline{\pi}_r$ by direct expression of the realised demands X_r in terms of the sum $(\Sigma_r x_{rs})$ of the individual outflows. Thus the combined entropy S_1 is expressed as the number of ways that the potential available supply capacity $\overline{X}_r$ may be allocated into a unutilised portion $(\overline{X}_r - \Sigma_s x_{rs})$ and utilised portions x_{rs} into each demand region s, yielding

$$S_1 = \pi_r [\, \overline{X}_r! / \{ (\overline{X}_r - \Sigma_s x_{rs})! \;\; \pi_s x_{rs}! \}] \tag{8.9}$$

where π here represents the product operation. If the log of (8.9) is taken and the Stirling approximation is applied in the usual way, the transformed entropy Z_1 comes out in the following form

$$Z_1 = -\Sigma_{rs} x_{rs} [\log x_{rs} - 1] - \Sigma_r (\overline{X}_r - \Sigma_s x_{rs}) [\log (\overline{X}_r - \Sigma_s x_{rs}) - 1] \tag{8.10}$$

The procedure is to maximize (8.10) in terms of x_{rs}, under constraints (8.7) associated with Lagrange multipliers λ_s and constraint (8.8) associated with Lagrange multiplier ν. This is seen to lead to the result

$$x_{rs} = \overline{X}_r \exp(\nu \pi_{rs} - \lambda_s) / [1 + \Sigma_s \exp(\nu \pi_{rs} - \lambda_s)] \tag{8.11}$$

Upon comparison with the relation (8.5) derived entirely from microeconomic arguments, we see that the results are identical when λ_s is simply interpreted as the scaled relation *(log 1/m)* $\Delta\tilde{\pi}_s$. *This compliance further reinforces the arguments in Chapter 3, where an alternative, purely microeconomic, definition of dispersion in the previous subsection reproduces the results of our entropy model, where the profit constraint, given as the objective of the non-dispersed (deterministic) theory, provides in conjunction with the entropy objective the identical dispersed result.*

8.2.2 A Model of Buyer/Consumer Behaviour

Distributional Approach. In the presence of link-specific fob prices p_{rs}, the corresponding cif buyer prices y_{rs} are given as

$$y_{rs} = p_{rs}(1 + t_{rs}) + c_{rs} \tag{8.12}$$

where t_{rs} is the assumed ad valorem tariff if the products cross national borders and c_{rs} are the unit transport plus handlings costs on link (r,s).

As emphasised in Roy and Johansson (1993), the key model enhancement required to handle differentiated goods and markets is to evaluate willingness-to-pay values w_{rs} in terms of the characteristics of the goods coming into region s from the alternative supply regions r. Thus, goods with high cif prices y_{rs} will still be in demand when the w_{rs} values are correspondingly high. Rather than expressing the model in relation to spillovers of demand to regions beyond that representing the minimum cif price suppliers, as in previous models, the spillovers are taken to be destined for regions beyond that having the maximum buyer/consumers surplus π_{rs}*, given as

$$\pi_{rs}^* = w_{rs} - y_{rs} \tag{8.13}$$

A log-linear model in terms of key characteristics of the goods (either intermediate goods or final demand) should be estimated on base period data to estimate the w_{rs} values, normalising such that the net surplus $[\Sigma_{rs} x_{rs}^{0} (w_{rs} - y_{rs}^{0})] = 0$.

As for the supply model, the first step is the specification of a demand share framework analogous to the supply share framework in (8.2). The log of the ratio at any demand region s of the amount x_{rs} demanded from a region r over the expected amount (D_s/n), where n is the number of supply regions, is expressed as a function of corresponding surplus differences in the form

$$log\ [\ x_{rs}/(D_s/n)] \ = \ \beta(\pi_{rs}* - \tilde{\pi}_s* - \Delta\tilde{\pi}_r*) \qquad (8.14)$$

where $\tilde{\pi}_s*$ is the perceived average surplus at s, $\Delta\tilde{\pi}_r*$ is a surplus correction factor to satisfy observed supplies from region r and β is a demand 'elasticity'. This form would allow a demand share model to be developed. However, the total regional demand D_s would still need to be supplied exogenously. Thus, a total short run regional import *capacity* $\overline{D}_s$ would need to be identified as a level of market saturation for consumer goods or intermediate goods in each region s. A relation analogous to (8.4) can be derived from the considerations in (8.3), finally yielding the result

$$log\ [\ D_s/(\overline{D}_s - D_s) \ = \ \beta\,\tilde{\pi}_s \qquad (8.15)$$

If the model (8.14)-(8.15) is estimated via Maximum Likelihood, as illustrated in Appendix 8.1 for the supply model, then the following equations emerge for the determination of the unknown parameters, Firstly, the parameters $\tilde{\pi}_s*$ are given via solution of the demand sum consistency conditions

$$\Sigma_r x_{rs} \ = \ D_s \qquad (8.16)$$

Secondly, the supply correction parameters $\Delta\tilde{\pi}_r*$ are obtained via constraining the model to produce the base period supplies in the form

$$\Sigma_s x_{rs} \ = \ X_r^{0} \qquad (8.17)$$

Finally, the parameter β is obtained by constraining the model to reproduce the base period surplus, expressed as

$$\Sigma_{rs}\ x_{rs}\ \pi_{rs}* \ = \ \Sigma_{rs}\ x_{rs}^{0}\ \pi_{rs}* \qquad (8.18)$$

Upon elimination of $\tilde{\pi}_r*$ in (8.14) and (8.15), the demand functions emerge as

$$x_{rs} \ = \ \overline{D}_s\ (1/n)\ exp\ \beta(\pi_{rs}* - \Delta\tilde{\pi}_r*)\ /\ [1 + (1/n)\ \Sigma_r\ exp\ \beta(\pi_{rs}* - \Delta\tilde{\pi}_r*)\ ] \quad (8.19)$$

This is seen to be a 'mirror image' of the supply function in (8.5). It again exhibits link-specific non-separability, with $\partial x_{rs}/\partial y_{us} \neq 0,\ u \neq r$. Also, the determination of 'elasticities' β_s for each demand region s would merely result in replacement of β by β_s in (8.19). With the demand model (8.19) containing the transport costs plus tariffs, it fits the conventional description of a spatial model. However, model (8.5) possesses these same properties purely through the link-specific prices p_{rs} and the link-specific costs $\bar{v}_{rs}$. Note that, if we assumed *uniform delivered* pricing rather than *fob* pricing, the situation would be reversed.

Entropy Approach. As the entropy derivation would be the 'mirror image of that already performed for the supply side in (8.9)-(8.11), it will not be spelled out here. The reader may take it as an exercise to carry out the entropy derivation, confirming that a result analogous to (8.19) is obtained. In fact, it is expressed as

$$x_{rs} = \bar{D}_s \, exp\,(\beta \pi_{rs}^* - \eta_r) / [\,1 + \Sigma_r \, exp\,(\beta \pi_{rs}^* - \eta_r)\,] \tag{8.20}$$

where η_r is the multiplier on the base period supply constraint (8.17).

8.2.3 Characterisation of the Equilibrium

Properties. Denoting producer flows with superscript p and buyer/consumer flows with superscript b, the link-specific prices p_{rs} must be associated with link-specific excess demand conditions on the notional excess demand $\tilde{Z}_{rs}$, given via

$$x_{rs}^b - x_{rs}^p = \tilde{Z}_{rs} \tag{8.21}$$

with x_{rs}^b expressed via (8.20) and x_{rs}^p via (8.11). Using a Walrasian approach (Intriligator, 1971), the corresponding equilibrium prices may, in principle, be determined by setting the excess demand $\tilde{Z}_{rs}$ equal to zero for each active link r,s.

In Roy and Johansson (1993), an existence proof based on just continuity and boundedness was provided for a producer/importer/consumer system with strong analogies with the above models (see Appendix 3.1). As an almost identical proof can be made for the above system, it will not be repeated here. However, the uniqueness and global stability proofs are described as follows.

First, the direct derivatives $\partial x_{rs}^b/\partial p_{rs}$ are given as

$$\partial x_{rs}^b/\partial p_{rs} = -\beta x_{rs}\,(1 + t_{rs})\,(1 - x_{rs} / \bar{D}_s) \tag{8.22}$$

Similarly for the cross-derivatives $\partial x_{rs} / \partial p_{us}$ and $\partial x_{rs} / \partial p_{rv}$ we have

$$\partial x_{rs}^b / \partial p_{us} = \beta x_{rs} x_{us}\,(1 + t_{us}) / \bar{D}_s \quad ; \quad \partial x_{rs}^b / \partial p_{rv} = 0 \tag{8.23}$$

With the demand function x_{rs}^{b} seen to be strictly decreasing in it's own price and strictly increasing in the cross-derivatives, gross substitutability is clearly satisfied (Intriligator, 1971). Turning to the supply model, we have

$$\partial x_{rs}^{p} / \partial p_{rs} = \nu x_{rs} (1 - x_{rs} / \overline{X}_{r}) \tag{8.24}$$

For the cross-derivatives, we obtain

$$\partial x_{rs}^{p} / \partial p_{us} = 0 \; ; \quad \partial x_{rs}^{p} / \partial p_{rv} = - \nu x_{rs} x_{rv} / \overline{X}_{r} \tag{8.25}$$

With strictly positive price derivatives for the own prices and negative cross-derivatives, gross substitutability is also satisfied on the supply side.

The next important point is to examine *global stability*. As shown in Intriligator (1971), this is assured if the excess demand Jacobian is diagonally dominant, that is, if we satisfy $|\partial \tilde{Z}_{rs} / \partial p_{rs}| - |\Sigma_{u \neq r} \partial \tilde{Z}_{rs} / \partial p_{us}| - |\Sigma_{v \neq s} \partial \tilde{Z}_{rs} / \partial p_{rv}| > 0$, where $| \ldots |$ represents the convention for 'absolute value'. Upon substitution from (8.21) to (8.25), this can be confirmed to yield

$$\beta x_{rs} [(1 + t_{rs}) - \Sigma_{u} x_{us} (1 + t_{us}) / \overline{D}_{s}] + \nu x_{rs} [1 - \Sigma_{v} x_{rv} / \overline{X}_{r}]$$

So long as there is some excess supply capacity $(1 - \Sigma_{v} x_{rv} / \overline{X}_{r})$ in each supply region r, the second term above is always positive. Considering the demand terms, so long as there is some unsatisfied demand in the system and the tariffs t_{us} into region s from alternative supply regions u do not vary too widely, then the first term is also positive, guaranteeing global stability.

Finally, following Roy and Johansson (1993), we examine the symmetry properties of the excess demand Jacobian. In fact, we lose symmetry defined in the form $\partial x_{rs}^{b} / \partial p_{us} \neq \partial x_{us}^{b} / \partial p_{rs}$ if the ad valorem tariffs into any regions s are *discriminatory*, that is, if $t_{us} \neq t_{rs}$ from any supply region u. In this case, the integrals for surplus evaluation are not path-independent and require identification of specific tâtonnement paths.

Solution Algorithm. As is discernible in the formulation of Roy and Johansson (1993), both the supply functions (8.11) and the demand functions (8.20) are invertible in terms of prices p_{rs}, that is, we have an explicit function expressing prices in terms of quantities. This means that the classical *cobweb* approach, as described in Intriligator (1971), may always be used to solve for the equilibrium. Alternatively, if (8.11) and (8.20) are substituted into (8.21), with the excess demands $\tilde{Z}_{rs}$ set to zero, the p_{rs} terms contained in each numerator can be combined and an iterative algorithm set up. For iteration t, this yields

$$p_{rs}^{t} = \{ 1 / [\nu + \beta (1 + t_{rs})] \} F_{rs}(\vec{p}_{uv}^{\,t-1}) \tag{8.26}$$

In the above, the F_{rs} terms are vector functions of all the prices p_{uv} as available at the previous iteration *(t-1)*. For the usual case of a diagonally-dominant Jacobian, convergence should be assured, except where highly discriminatory tariffs cause serious loss of symmetry. Note further, that if links having zero base period flows are considered also to have zero flows in the future, the special information theory approach in Roy (1987) [see also Sect. 1.3.1] should be applied to adjust both (8.11) and (8.20).

8.3 Consideration of Imperfect Competition

Many explanations have been sought to explain the rather inconsistent empirical performance of the classical SPE models. The introduction of product heterogeneity, as well as dispersion of destination flows beyond that from the minimum cif price source represented in the DSPE models discussed earlier, has the potential to provide improved matching with observed flows and pricing. Even this may not be sufficient. As discussed by Kolstad and Mathiesen (1991), the perfect competition paradigm has been under challenge for some time in these classes of models. Even for bulk commodities, such as oil, natural gas and wheat, oligopolistic conduct is clearly observable. Similar behaviour obviously exists in the markets for more differentiated goods, such as automobiles. Although these doubts about the perfect competition models have existed for some time, implementation of imperfect competition has foundered on computational difficulties, as well as on the variety of assumptions to be made about collusion and conjectural variations (Intriligator, 1971). There are also examples of predatory behaviour, which are very difficult to model within the usual imperfect competition paradigms. Nevertheless, introduction of Cournot-Nash assumptions into the classical SPE framework, including solution via variational inequalities, has been achieved by Miyagi (1991), Kolstad and Mathiesen (1991) and Nagurney *et al.*(1994). In the following, imperfect competition is introduced into the DSPE class of model described in the previous section. In conjunction with our DSPE framework, we do not need imperfect competition to better 'fill out' unrealistic zero entries in the flow matrix, as occurs for the SPE models. Instead, the aim is to introduce oligopoly behaviour for the regions or countries which appear to dominate the market. The framework is developed as a Cournot quantity model, augmented by illustration of the evaluation of conjectural variations within a Stackelberg leader-follower procedure.

8.3.1 Specification of an Equilibrium

As in Kolstad and Mathiesen (1991), it is assumed that the producing regions r can be divided into two groups, the oligopolists 1 to f and the price-takers (the 'competitive fringe') $f+1$ to n. Examining first a Cournot quantity model

(Intriligator, 1971), it is necessary to invert the demand functions (8.21) to express the prices p_{rs} as functions of the quantities x_{us}, $u = 1,m$. Using symbolic matrix algebra, it can be shown that this can be achieved in the elegant closed form

$$p_{rs}^{b} = \{ - \log [x_{rs} \exp - \beta(w_{rs} - c_{rs} + \eta_r) / (\overline{D}_s - \Sigma_u x_{us})] \} / \beta(1 + t_{rs}) \quad (8.27)$$

This function must be inserted in place of p_{rs} in (8.1) and (8.8) and included in the entropy maximisation procedure (8.9) to (8.11). Omitting, à la Cournot, the effects of changes $\partial p_{us}^{b} / \partial x_{rs}$ in the prices p_{us} offered by region s to producers in other regions u $(u \neq r)$ caused by changes in the amount x_{rs} supplied by r to s, the perfectly competitive supply function in (8.11) for the oligopolists $r = 1,f$ is modified to[1]

$$x_{rs} = \overline{X}_r \exp \nu(p_{rs}^{b} - \overline{v}_{rs} - \lambda_s + \partial p_{rs}^{b} / \partial x_{rs}) / [\, 1 + \Sigma_s \exp \nu(p_{rs}^{b} - \overline{v}_{rs} - \lambda_s + \partial p_{rs}^{b} / \partial x_{rs}) \,] \quad (8.28)$$

where both p_{rs}^{b} and $\partial p_{rs}^{b} / \partial x_{rs}$ are themselves functions of x_{us}, $u = 1,m$, as seen in (8.27) and it's derivative

$$\partial p_{rs}^{b} / \partial x_{rs} = -\{ [1/x_{rs}] + [1/(\overline{D}_s - \Sigma_u x_{us})] \} / \beta(1 + t_{rs}) \quad (8.29)$$

As (8.28) already includes the inverse demand functions p_{rs}^{b}, its solution for x_{rs} may directly yield the equilibrium flows, from which the corresponding equilibrium prices for the oligopolists are obtained via (8.27). With existence criteria similar in principle to those given in Roy and Johansson (1993) for competitive equilibria, we just need to examine the uniqueness properties of (8.28) as follows.

Proposition 8.1 *Using the inverse demand functions of (8.27) and the alternative supply functions (8.11) and (8.28) of the combined oligopoly/competitive system, the resulting equilibrium is unique.*

Proof: From (8.29), the second derivative of the inverse demand function is

$$\partial^2 p_{rs}^{b} / \partial x_{rs}^{2} = \{ [1/x_{rs}^{2}] - [1/(\overline{D}_s - \Sigma_u x_{us})] \} / \beta(1 + t_{rs})$$

We consider the entropy objective (8.10) of the supply function Z' enhanced with the right-hand sides of the constraints and their associated Lagrange multipliers, expressed as the 'equivalent' unconstrained objective

[1] In the spirit of oligopolists having individual regional profit functions, the profit parameter for regions $r = 1,f$ should become the regional-specific value ν_r. But for monopolistic competition, the single value ν should be retained.

$$Z' = -\sum_{rs} x_{rs} [\log x_{rs} + \lambda_s - \beta(p_{rs} - \bar{v}_{rs}) - 1] - \sum_r (\bar{X}_r - \sum_s x_{rs}) [\log (\bar{X}_r - \sum_s x_{rs}) - 1]$$

Then we can write

$$\partial Z'^2 / \partial x_{rs}^2 = -\{ [1/x_{rs}] + [1/(\bar{X}_r - \sum_s x_{rs})] - v [2\, \partial p_{rs}^b / \partial x_{rs} + x_{rs}\, \partial^2 p_{rs}^b / \partial x_{rs}^2] \}$$

Upon substitution from (8.29) and the second derivative above, we have

$$\partial^2 Z' / \partial x_{rs}^2 = -\{ [1/x_{rs}] + [1/(\bar{X}_r - \sum_s x_{rs}) + v [(1/x_{rs}) + 2/(\bar{D}_s - \sum_r x_{rs}) + x_{rs}/(\bar{D}_s - \sum_r x_{rs})^2] / \beta (1 + t_{rs}) \}$$

Similarly, the cross-derivatives come out as

$$\partial^2 Z' / \partial x_{rs}\, \partial x_{us} = v\{ [1/(\bar{D}_s - \sum_r x_{rs})] + x_{rs} / (\bar{D}_s - \sum_r x_{rs})^2] \} / \beta (1 + t_{rs})$$

as well as

$$\partial^2 Z' / \partial x_{rs}\, \partial x_{rv} = -1 / (\bar{X}_r - \sum_s x_{rs})$$

Thus, the direct second derivatives of the maximization problem are strictly negative and all the principal minors of the second derivative determinant are strictly positive. This guarantees a unique solution for the oligopoly equilibrium. Also, by setting $\partial p_{rs}^b / \partial x_{rs}$ and the higher derivatives to zero in the above, it can be seen that the same conditions are satisfied for the competitive fringe. Thus, the overall equilibrium is unique. *Q.E.D.*

8.3.2 Elements of a Stackelberg Approach

In the Cournot quantity model above, each oligopolist is taken only to consider the price reactions of the demand markets to it's quantity decisions, neglecting direct reaction by its fellow oligopolists or direct reactions to expected behaviour of its fellow oligopolists. In oligopoly theory (Intriligator, 1971), this is characterised as a case of *zero conjectural variation*, that is $\partial x_{rs} / \partial x_{us} = 0$ from the viewpoint of region r with respect to region u. Under this assumption, the result of setting the quantity derivative $\partial Z' / \partial x_{rs}$ of the unconstrained Lagrangian Z' above to zero under this assumption leads to

$$\partial Z' / \partial x_{rs} = -\log x_{rs} + \log (\bar{X}_r - \sum_s x_{rs}) + v (p_{rs} - \bar{v}_{rs} + x_{rs}\, \partial p_{rs} / \partial x_{rs} = 0 \quad (8.30)$$

Denoting the above derivative as F_{rs}, we can use implicit derivation $\partial x_{us} / \partial x_{rs} = -(\partial F_{rs} / \partial x_{rs}) / (\partial F_{rs} / \partial x_{us})$ to express these extra effects as

$$\partial x_{us} / \partial x_{rs} = \{ (1/x_{rs}) + 1/(\overline{X}_r - \Sigma_s x_{rs}) - v(2\, \partial p_{rs} / \partial x_{rs} + x_{rs}\, \partial^2 p_{rs} / \partial x_{rs}^2) \} \\ / \{ v(\partial p_{rs} / \partial x_{us}) + x_{rs}\, \partial^2 p_{rs} / \partial x_{rs}\, \partial x_{us}) \} \quad (8.31)$$

Upon using the derivatives from the previous section this finally yields

$$\partial x_{us} / \partial x_{rs} = \{ \beta(1 + t_{rs}) [(1/x_{rs}) + 1/(\overline{X}_r - \Sigma_s x_{rs})] + v[(1/x_{rs}) + 2/(\overline{D}_s - \Sigma_r x_{rs}) \\ + x_{rs}/(\overline{D}_s - \Sigma_r x_{rs})^2] \} / v\{ 1/(\overline{D}_s - \Sigma_r x_{rs}) + x_{rs}/(\overline{D}_s - \Sigma_r x_{rs})^2 \} \quad (8.32)$$

Thus, if region r believes, à la Stackelberg, that region u will act as a Cournot oligopolist, the term $x_{rs}\, (\partial p_{rs} / \partial x_{us})\, (\partial x_{us} / x_{rs})$ should be added to $x_{rs}\, \partial p_{rs} / \partial x_{rs}$ in (8.29). From comparison of (8.29) with (8.32), it is clear that this enhancement does not materially add to the computational complexity. However, the main difficulty arises in the assignment of Cournot or Stackelberg behaviour from each oligopolist with respect to each other oligopolist and the potential occurrence of 'Prisoners' Dilemma' situations (Intriligator, 1971). Thus, in practice, this extension would only appear to be feasible in markets dominated by two or three major players. Such ambiguity is endemic in modelling imperfect competition.

8.3.3 Elements of a Solution Procedure

Upon substitution of (8.27) and (8.29) into (8.28), a relationship is given entirely in terms of x_{rs}, with $(\overline{D}_s - \Sigma_r x_{rs})$ and the Σ_s terms in the denominator of (8.28) providing the spatial interdependencies. The clearest approach is clearly to start with a feasible solution x_{rs}^1, perhaps representing an appropriate updating of the base period flows x_{rs}^0, evaluating (8.27) and (8.29) to update (8.28) and using (8.11) for the competitive fringe. This would proceed until $x_{rs}^t \approx x_{rs}^{t-1}$ to an acceptable accuracy for all (r,s) combinations. Final substitution into (8.28) would yield the corresponding prices. Convergence of this simple procedure would need to be checked both mathematically and empirically. Of course, we can always turn to the Newton-Raphson procedure in an emergency. With the proof of the negative-definiteness of the Jacobian in *Proposition 8.1*, stable solution procedures are to be expected. Also, the above developments illustrate that plausible supply and demand functions possess convenient properties such as analytical invertibility. This opens up the area for efficient empirical testing with respect to various behavioural assumptions within the imperfect competition paradigm.

8.3.4 Surplus Evaluation

The evaluation of alternative trade strategies requires the regional evaluation of changes in both producer and consumer surplus in the before ('0') and the after ('1') situations. As shown in the important paper by Champernowne *et al.* (1976) and discussed in Sect. 3.2.8, the surplus corresponds to the transformed unconstrained objective of the maximization problem divided by the Lagrange multiplier of the economic constraint. Thus, the change ΔP_r of the producer surplus in region r between a future period '1' and the base period '0' comes out via the Z' expression in *Proposition 8.1* in the form

$$\Delta P_r = \{ [(\Sigma_s x_{rs}^1 p_{rs}^1) - K_r^1] - [(\Sigma_s x_{rs}^0 p_{rs}^0) - K_r^0] \} \tag{8.33}$$

where the non-linear cost functions K_r^t at any stage t (0 or 1) come out as

$$K_r^t = \Sigma_s x_{rs}^t (\bar{v}_{rs}^t + \lambda_s) + \{ \Sigma_s x_{rs}^t (\log x_{rs}^t - 1) + (\bar{X}_r^t - \Sigma_s x_{rs}^t) [\log(\bar{X}_r^t - \Sigma_s x_{rs}^t) - 1] \} / v \tag{8.34}$$

From, the above, it can seen that expected costs are high and producer surplus relatively low for small values of the 'elasticity' v. A completely analogous derivation can be made for buyer/consumer surplus.

8.4 Future Directions

The preceding formulation, both for perfect and imperfect competition, has concentrated on transport of the outputs, absorbing the cost of the inputs and their transport costs into the cost term $\bar{c}_{rs}$. By avoidance of explicit identification of the quantity and sectoral composition of the inputs, a production function has become superfluous. A future challenge is to generalise the formulations to include these above features, as well as to include production scale economies at the firm level. In order to gain guidance on the introduction of production functions into the analysis, we can be influenced by the producer model in Sect. 3.2.3. As far as handling an explicit mix of inputs is concerned, we can be guided by both Sects. 7.1 and 7.2. Scale economies at the firm level are elegantly handled following the approach of Fujita *et al.* (1999), as summarised in Sect. 2.3.1. Thus, we have the ingredients to formulate more realistic models, both for perfect and imperfect competition. However, especially for imperfect competition, empirical verification will still be difficult, even with the more complete formulation sketched above. Alternative leader/follower scenarios will need to be tested, as well as the possibilities for predatory behaviour.

Appendix 8.1

In Appendix 1.1, maximum likelihood estimation was illustrated for the floorspace parameter of the classical retail model. We illustrate the procedure here for the estimation of the parameter $\Delta\tilde{\pi}_s$ in the producer supply model (8.5).

We see that not only are the product flows x_{rs} results of this model, but also the unutilised capacity $(\overline{X}_r - X_r)$ in each producing region r. Thus the maximum likelihood criterion must be expanded accordingly, involving maximization of the following objective Y in terms of the unknown parameters

$$Y = \Sigma_{rs}\, x_{rs}^0 \log\{x_{rs}/[\Sigma_{rs}\, x_{rs} + \Sigma_r(\overline{X}_r - X_r)]\} + \Sigma_r(\overline{X}_r - X_r^0)\log\{(\overline{X}_r - X_r) \\ /[\Sigma_{rs}\, x_{rs} + \Sigma_r(\overline{X}_r - X_r)]\}$$

where items with zero superscripts relate to base period observations and the rest refer to the model. From the discussion around (8.5), $(\overline{X}_r - X_r)$ can be expressed as

$$(\overline{X}_r - X_r) = \overline{X}_r/[1 + (1/m)\,\Sigma_s \exp v(\pi_{rs} - \Delta\tilde{\pi}_s)].$$

Together with this plus (8.5), Y can be written as

$$Y = \Sigma_{rs}\, x_{rs}^0\{\log(\overline{X}_r/m) + v(\pi_{rs} - \Delta\tilde{\pi}_s) - \log[1 + (1/m)\,\Sigma_s \exp v(\pi_{rs} - \Delta\tilde{\pi}_s)]\} \\ + (\Sigma_{rs}\, x_{rs}^0)\log\{\Sigma_{rs}(\overline{X}_r/m)\exp v(\pi_{rs} - \Delta\tilde{\pi}_s)/[1 + (1/m)\,\Sigma_s \exp v(\pi_{rs} \\ - \Delta\tilde{\pi}_s)] + \Sigma_r\, \overline{X}_r/[1 + (1/m)\,\Sigma_s \exp v(\pi_{rs} - \Delta\tilde{\pi}_s)]\} + \Sigma_r(\overline{X}_r - X_r^0) \\ \{\log \overline{X}_r - \log[1 + (1/m)\,\Sigma_s \exp v(\pi_{rs} - \Delta\tilde{\pi}_s)]\} + [\Sigma_r(\overline{X}_r - X_r^0)]\log\{ \\ \Sigma_{rs}(\overline{X}_r/m)\exp v(\pi_{rs} - \Delta\tilde{\pi}_s)/[1 + (1/m)\,\Sigma_s \exp v(\pi_{rs} - \Delta\tilde{\pi}_s)] + \Sigma_r\, \overline{X}_r/ \\ [1 + (1/m)\,\Sigma_s \exp v(\pi_{rs} - \Delta\tilde{\pi}_s)]\}$$

Before proceeding further, it can be seen that the identical set of terms within the second and fourth set of bold $\{\ldots.\}$ brackets reduces to the constant term $(\Sigma_r\, \overline{X}_r)$. This simplifies the following algebra significantly. Then we write the condition for maximization in terms of $\Delta\tilde{\pi}_s$ as

$$\partial Y/\partial \Delta\tilde{\pi}_s = -v\{\Sigma_r\, x_{rs}^0 + \Sigma_r(\Sigma_s\, x_{rs}^0)(1/m)\exp v(\pi_{rs} - \Delta\tilde{\pi}_s)/[1 + (1/m) \\ \Sigma_s \exp v(\pi_{rs} - \Delta\tilde{\pi}_s)] + \Sigma_r(\overline{X}_r - X_r^0)(1/m)\exp v(\pi_{rs} - \Delta\tilde{\pi}_s) \\ /[1 + (1/m)\,\Sigma_s \exp v(\pi_{rs} - \Delta\tilde{\pi}_s)]\} = 0$$

Upon noting that $\Sigma_s\, x_{rs}^0 = X_r^0$, the above simplifies to

$$\Sigma_r \overline{X}_r (1/m) \exp \nu (\pi_{rs} - \Delta\tilde{\pi}_s) / [1 + (1/m) \Sigma_s \exp \nu (\pi_{rs} - \Delta\tilde{\pi}_s)] = \Sigma_r x_{rs}^0$$

Upon realising via (8.5) that the term inside the summation sign on the left-hand side is simply x_{rs}, the condition for determination of $\Delta\tilde{\pi}_s$ becomes

$$\Sigma_r x_{rs} = \Sigma_r x_{rs}^0$$

This implies that the model must preserve the observed column sums of the flow matrix. These conditions, quoted in (8.7), correspond precisely to the constraints imposed on the equivalent entropy model of (8.11). A similar derivation based on the parameter ν can be shown to yield the profit constraint relation (8.8).

9 Interregional Migration

9.1 Introduction

The movement of human beings on the Earth has been a characteristic of the human race since the dawn of civilisation. In the early stages, most of the movement was *nomadic*, where people had no fixed place of abode. In some cases, there was one more or less fixed location for the winter period and another for the summer. With the growth of cities in the first ancient civilisations, we first entered the realm of *internal interregional migration*, the main focus of this chapter. In other words, we started to have people making permanent or semi-permanent moves from the country to the city and occasionally vice-versa, as well as having people moving between cities, within the one culture. Whereas the reasons for these moves were often economic, they often related to other factors, such as escape from battle areas, availability of fresh water and preference for defensible locations.

Modern interregional migration is clearly also influenced by economic factors. On the other hand, with basic utilities such as power, water supply and sewerage being available in most locations in First World countries, other factors come into play. These often relate to *lifestyle* choices, including equable climate, air quality, reduced congestion, quality of nearby educational facilities, health care access and entertainment options.

In this introduction, some key issues in the understanding and modelling of interregional migration are highlighted, followed by a summary of related work. This is followed by formulation of a comparative static model of interregional migration as introduced by Roy and Flood (1992) as part of a national study in Australia (Flood *et al.* 1991). Finally, we postulate a two-stage dynamic migration model which defines the stages as in Weidlich and Haag (1988), but presents a fundamentally different interaction model.

9.1.1 Market Segmentation

It is clear that the decision to migrate of a young single man in his early twenties may have very different motivations to that of a recently retired married couple. Thus, we must at least segment the market according to age of migrant, carrying out a separate estimation for each segment. In this process, we need to consider a degree of segmentation which still ensures that the migration flows of that segment between any two regions are not too small for statistical purposes. Nevertheless, we should often try to segment more finely than just by age. For example, some other segmentation criteria include gender type, income category, job skill level (unskilled, semi-skilled or skilled), tenure status of household (renter or owner-occupier), employment status and household type, including numbers of children. The type of segmentation also clearly depends on the purpose of the study for which the model is to be applied. Those criteria of lesser importance for a particular study can still be catered for by introducing them as dummy variables within the model. Thus, for employment status, we may set a dummy variable of zero for unemployed migrants and a value of unity otherwise. The most appropriate level of market segmentation for a particular study is based on judgment acquired from previous studies.

9.1.2 Choice Hierarchy

It is useful to consider two stages in a migration decision, the overall decision to move and the choice of destination. These two steps may not necessarily be sequential. In some cases, one may have one highly preferred destination, but may still be deciding upon whether or not to move. In other cases, one may suddenly want to leave where one is, without yet any clear idea on where to go. Thus, the choices are best considered in some sort of interdependency framework, either as simultaneous or nested sequential. For example, Odland (1988) developed a sequential nested logit model of the decision whether or not to migrate together with the choice of destination, given that one is a mover. In principle, such a structure can also be handled by the 'competing destinations' model framework introduced by Fotheringham (1983a) and discussed in Sect. 1.3.5. Alternatively, a nested entropy-type model, such as that introduced by Roy and Lesse (1985) and discussed in Sect. 3.2.5, may be usefully applied. The latter has the structure of the more sophisticated FIML (Full Information Maximum Likelihood) nested logit model described by McFadden (1981). The model generates a potentially different *weighting* on the entropy of the decision to move compared with that of choice of destination for the movers.

9.1.3 Reliability of Projections

Earlier gravity-type migration models can produce errors in flows of 40-50%, as discussed by Plane (1982). This weakness is primarily due to the fact that the distance of the move is only one factor influencing migration. It is likely that if other plausible influences such as wage rates, housing prices and unemployment rates were included as additional explanatory factors in the analysis, these errors could be reduced. As demonstrated further by Plane (1982), information-theoretic approaches, even in their simplest form, could reduce such flow errors to 10-15%.

As discussed in Sect. 1.3.1, the choice of an entropy model vs a pure information theory model is not always obvious. Whereas the former allow changes in the explanatory variables between the base period and the projection period to influence the model results, additional qualitative factors are ignored. On the other hand, as information theory approaches implicitly include, but do not individually identify, the explanatory factors, they cannot handle the influences on the flow projections of changes in these factors. This 'paradox' was investigated in Roy (1987), where a hybrid approach was presented to address the above difficulties. It was formally shown that if an entropy-type model, with all the behavioural constraints attached, were generalised to an information theory model by applying appropriate endogenously-determined prior probabilities, it would (i) yield identical Lagrange multipliers to those of the original entropy model where the prior probabilities are uniform or non-existent and (ii) produce, when applied at the base period a set of modelled flows *identical* to those observed. Thus, we have produced a resolution of the paradox, with the explanatory variables and their associated Lagrange multipliers being present to allow future variable changes to be introduced as for entropy models, but with the computed prior probabilities still being available to be applied to reflect the influence of unexplained factors. Such an approach is used here. Following Rogerson and Plane (1984), the *base period* flows used as the basis for the projections need not correspond to the most recent data available, but may be a weighted average of the normalised flows from previous periods. The best weightings may be determined by experimentation on the chosen set of normalised prior flow matrices in order to obtain the best projections of the flows for the most recent data set. This process will allow an appropriate *discounting rule* to be developed to reflect the influence of the results of older data sets on projecting future flows.

9.1.4 Net Migration vs Gross Migration Models

As demonstrated by the well-respected demographer Rogers (1966), the steady-state growth solution of a Markov-type migration model with the transition probabilities constant over time yields the in-migration rate for each region equal to the out-migration rate plus a small local correction. This argument was further expanded by Mueser and White (1989). Thus, equilibrium is characterised by *zero net migration*. Net migration models dominated the earlier literature, particularly in economics, because of the concern of policymakers with net population changes in estimating future infrastructure requirements. However, from a behavioural perspective, the *net migrant* is a non-identifiable abstraction (Rogers, 1989). In addition, estimates of net migration provide no idea of the *mobility (i.e. turnover)* of the population, which can sometimes be a measure of the dynamism of cities and regions. Thus, more recent work in regional science concentrates on modelling *gross* migration, distinguishing the in-flows from the out-flows. Of course, the net flows are immediately available as differences of the gross flows, but not vice-versa. This approach is reflected in the models of Plane and his colleagues mentioned earlier, as well as in later work in Sweden carried out by Holmberg (1987). The current chapter is devoted to modelling gross migration.

9.1.5 Exogenous and Endogenous Quantities

In most of the earlier models of migration, such as in Plane (1982), it was assumed that the in-migration totals and the out-migration totals over the entire regional system were somehow available exogenously at the projection period; the task of the model then became the allocation of these flow totals to flows between the different regional pairs. However, as these flow totals express *mobility*, influenced by the same economic/demographic factors included in the model to allocate the interregional flows, it is not reasonable to expect them to be magically provided external to the model. In later work (Plane and Rogerson, 1985), this weakness was overcome, and proportional adjustment techniques reconciled the gravity and demographic frameworks, whereas shift and share analysis (Plane, 1987) allowed changes in flows to be composed into components which were easy to interpret. In this chapter, a more formal approach is applied to overcome this difficulty, as illustrated in Sect. 3.2.4. After the model is estimated on various economic and population data at the base period, including the overall total of in-flows and out-flows, invariant Legèndre transforms of the constrained entropy 'objective', plus the corresponding transformation of the associated Lagrange multipliers to parameters, allows the regional in-flow and out-flow totals to become endogenous output in the projection periods. Of course, these Lagrange multipliers as overall propensities to in-migrate or out-migrate will not remain constant over time. By a series of estimations at successive earlier time periods, one may be able to identify dynamic trends in these parameters, thus permitting them to be updated for use in future projection periods. In other words, following Wilson (1970), the model is estimated as doubly-constrained, with the transformations for projection resulting in just an origin-constrained version. We merely require the evolving regional population totals at the origins rather than exogenous estimates of the regional out-migration totals.

9.1.6 Updating of Regional Populations

When we are looking at movements over discrete time periods, we need to adopt a scheme for updating populations. In fact, following Holmberg (1990), we use the well-known *demographic accounting principle*. Firstly, we must distinguish between the movers and the stayers in each region. The interregional migration flows are arranged in a square matrix of dimension equal to the number of regions, whereby the diagonal terms in any row or column represent the stayers in that region. The row sums are taken to represent the population in each region at the *start* of the time period, made up of the sum of all the outflows plus the stayers. On the other hand, the column sums represent the population in each region at the *end* of the time period, as the sum of stayers plus inflows from all other regions. With the use of this convention plus the opening of the model described above, we have established a framework for a population change model, rather than a migration model *per se*. The system becomes responsive to the influences of both economic and demographic change.

The development of such a population change model permits the behavioural constraints to be expressed in two alternative forms. One possibility is that the model is constrained to reproduce base period average conditions, such as unemployment rates averaged over the set of regions. Alternatively, we may choose to use projected average *changes* of unemployment rates over the time period as explanatory variables influencing migration.

Finally, the product form of most migration models in terms of origin push and destination pull implies that the volume of migration increases according to increase in the products of the origin and destination population totals. For example, if the population in each of two regions increased by 10%, the model would project a $100[1.1^2 - 1^2] = 21\%$ increase in migration. Models should be formulated with the flexibility to test this hypothesis, already challenged by Dorigo and Tobler (1983). It would be plausible to expect actual behaviour to lie somewhere between a linear increase (10%) and a quadratic increase (21%).

9.1.7 Properties of the Migration Flow Pattern

Most spatial interaction models yield *transitive* flows, as pointed out by Goodchild and Smith (1980). In other words, if M_{ij} represents the number of migrants moving from region i to region j, then for a third region k, *transitive* flows are defined when if $M_{ij}>M_{ji}$ and $M_{jk}>M_{kj}$, then $M_{ik}>M_{ki}$. Qualitatively, we may state the transitivity condition thus: *When region i is a net exporter of migrants to region j, and region j a net exporter of migrants to a third region k, then region i is also a net exporter of migrants to region k.* Despite the plausibility of this condition, observed migration flows in the United States exhibit considerable intransitivities (Goodchild and Smith, 1980). Similar properties are to be observed in recent migration flows within Australia. The implication of this is that the model structure must be sufficiently flexible to handle such flow patterns. This is achieved in the model to be described below. Such flow pattern properties can result from inherently unsymmetrical preferences, such as greater tendencies for older migrants to move from cool to warm climates rather than vice-versa.

9.2 Comparative Static Migration Model

As discussed in Sect. 9.1.1, the population is segmented into a certain number of groups which can be regarded as relatively homogeneous with respect to the aims of the study being undertaken. The model is then estimated separately for each such group. Interactions between the groups then occur merely via cohort transitions due to aging. The formulation below is for one particular group k.

9.2.1 Notation

Let all the base period observed quantities be denoted with a superscript '*0*'. These may be an appropriate weighted sum of data for two or more of the most recent time periods, as suggested by Rogerson and Plane (1984). Our objective is to determine the number of migrants M_{ij} moving from region i to region j during a specified time period. The diagonal term M_{ii} refers to the number of stayers, including intraregional movers, within region i, who are lumped together by the model. Let P_i^0 be the population in region i at the start of the base time period. Also, let U_i^0 be the number of emigrants leaving region i to proceed out of the regional system during the base period, I_i^0 be the number of immigrants coming into region i from outside the system, D_i^0 be the number of deaths in region i during the base period and B_i^0 be the corresponding number of births. This demographic data needs to be supplied exogenously for the base period. Let M_j^0 be the observed number of migrants arriving in region j during the base period and M^0 the total number of base period migrants in the system.

Coming to behavioural data, let c_{ij}^0 be the average travel cost of a move between region i and region j and c^0 be the observed average travel cost of all migrants. Further behavioural data includes origin-based items affecting the decision of whether or not to move, as well as information on expected gains (or losses) upon moving to different regions. Different behavioural factors will be relevant for the different age/lifestyle groups k. For example, knowing the regional unemployment rates r_i^0 in each region i, the parameter associated with the average rate r^0 over the entire system is one measure of the disincentive to move. Note, however, this influence may be irrelevant for migration moves of retirees - in fact, high unemployment rates in a region may often be an incentive for retirees seeking peace and quiet and associated lower housing costs. On the other hand, when looking at destinations of potential movers, housing cost differentials $(h_i^0 - h_j^0)$ and unemployment rate differentials $(r_i^0 - r_j^0)$ could have significant influences on regional choice. However, if the high rents in a region are more than compensated for by its high amenity, they are likely to have a positive impact on choice. If we apply constraints for such factors, their right-hand sides reflect average rates of change over the entire regional system during the base period. Thus, we can constrain the model to reproduce the average loss in unemployment r^{l0} and the average reduction of housing costs h^{r0} by the movers during the base period. Also, in looking at the 'sunbelt' phenomenon in countries like the US and Australia, we may define dummy variables n_{ij} as unity if the move from i to j represents a move from a cool region i to a warm region j, and zero otherwise, given an average n^0. This constraint can pick up intransitivities in the flows. Finally, in order to give the model the non-IIA property reflecting different uncertainties in the decision to move vs the choice of alternative destinations, we constrain the model to reproduce the observed entropy S^0 of the total out-movers plus the stayers, as discussed in principle in Sect. 1.3.4. This allows the uncertainty of the overall decision of whether or not to move to be replicated by the model, introducing a nested structure similar to nested logit.

9.2.2 Model Estimation

Entropy Procedure. Consider the entropy S^0 of the movers and the stayers to be maximised at the base period. This is the simple flow entropy

$$S^0 = -\Sigma_{ij} M_{ij} [\log M_{ij} - 1] \tag{9.1}$$

where the stayers are represented in the diagonal (i,i) terms. Using the demographic accounting principle discussed earlier, we relate in each region i the base period population net of emigrants and deaths to the out-movers and stayers via the following constraints

$$M_{ii} + \Sigma_{j \neq i} M_{ij} = P_i^0 - U_i^0 - D_i^0 \tag{9.2}$$

Also, we want to ensure that the model reproduces the total number of in-movers M_j^0 into each region j via the constraints

$$\Sigma_{i \neq j} M_{ij} = M_j^0 \tag{9.3}$$

Then, the average travel costs are reproduced via the constraint

$$\Sigma_{ij(\neq i)} M_{ij}\, c_{ij}^0 = c^0 M^0 \tag{9.4}$$

Regarding the overall decision to move, we can constrain the model to represent the average unemployment rate of the movers via the constraints

$$\Sigma_{ij(\neq i)} r_i^0 M_{ij} = r^0 M^0 \tag{9.5}$$

Note that the above constraint strictly implies that we can obtain the regional unemployment rates of the movers! These could only be made available via a special survey. Otherwise, if we are to use overall regional unemployment rates, we are assuming that the relatively small cohort of movers are in this sense representative of the population as a whole. This assumption could clearly be a source of error, as stressed by Van Dijk and Oosterhaven (1986). Similar caveats apply to the data associated with other behavioural variables. We now consider the cool to warm climate phenomenon via

$$\Sigma_{ij(\neq i)} M_{ij}\, n_{ij} = n^0 M^0 \tag{9.6}$$

Considering the potential reduction of unemployment rates experienced by the movers, we apply the following constraint

$$\Sigma_{ij(\neq i)} M_{ij} (r_i^0 - r_j^0) = r^{10} M^0 \tag{9.7}$$

Similarly, the reduction in housing rents of the movers is given by

$$\Sigma_{ij(\neq i)} M_{ij} (h_i^0 - h_j^0) = h^{r0} M^0 \tag{9.8}$$

Note, there is no guarantee that the signs on the right-hand sides of (9.7) and (9.8) will be positive. For example, young migrants may deliberately move into regions with higher rents to obtain improved job prospects. Conversely, retirees on welfare may deliberately choose to move to low cost areas, which may often also have higher unemployment. As a result, to avoid too much cancellation of opposing effects in these constraints, we may need to have a separate constraint for highly rural destination regions and another for the more urbanised regions.

Finally, we come to the pivotal constraint on the entropy of the sum of stayers and out-movers from a region, expressed as

$$- \Sigma_i (\Sigma_{j\neq i} M_{ij}) [\log (\Sigma_{j\neq i} M_{ij}) - 1] - M_{ii} [\log M_{ii} - 1] = S^0 \tag{9.9}$$

We could clearly devise other behavioural constraints, depending on the age/lifecycle group being modelled. For example, in Australia, coastal regions have a special amenity attraction, which should be accounted for in the model. Also, regions containing a large capital city should be singled out by the use of a dummy variable.

Now, if the entropy in (9.1) is maximised in terms of M_{ij} under constraints (9.2) to (9.9), we obtain the most probable flows. Let (9.2) be associated with Lagrange multipliers λ_i^0, (9.3) with ψ_j, (9.4) with β, (9.5) with η, (9.6) with ϕ, (9.7) with $-\gamma$, (9.8) with $-\alpha$ and (9.9) with $-(1-\theta)/\theta$. The special form of the latter multiplier is chosen to make the analogies with the nested logit model more transparent. Letting $M_i = \Sigma_{j\neq i} M_{ij}$, the total number of migrants leaving region i, $A_i = exp - \lambda_i$, $B_j^0 = exp - \psi_j$, and $f_{ij}^0 = exp [\alpha (h_i^0 - h_j^0) + \gamma (r_i^0 - r_j^0) - \beta c_{ij}^0 - \eta r_i^0 + \phi n_{ij}]$, we obtain for the interregional flows M_{ij}

$$M_{ij} / M_i = B_j^0 f_{ij}^0 / [\Sigma_{j\neq i} B_j^0 f_{ij}^0) \tag{9.10}$$

This represents the conditional probability of choosing destination region j given that one has decided to move from region i. For the stayers, we have simply

$$M_{ii} = A_i^\theta \tag{9.11}$$

The out-movers M_i from region i are expressed via

$$M_i = A_i^\theta [\Sigma_{j\neq i} B_j^0 f_{ij}^0]^\theta \tag{9.12}$$

Letting A_i^θ be denoted by $\overline{A}_i$, the substitution of (9.11) and (9.12) into the net population constraint (9.2) yields

$$\overline{A}_i = (P_i^0 - U_i^0 - D_i^0) [1 + (\Sigma_{j\neq i} B_j^0 f_{ij}^0)^\theta]^{-1} \tag{9.13}$$

Expressing an intermediate variable $\overline{C}_i$ as

$$\overline{C}_i = [\Sigma_{j\neq i} B_j^0 f_{ij}^0]^{\theta-1} \tag{9.14}$$

the destination factors B_j^0 are determined recursively with the origin factors $\overline{A}_i$ in (9.13) and $\overline{C}_i$ in (9.14) via the relations

$$B_j^0 = M_j^0 [\Sigma_{i\neq j} \overline{A}_i \overline{C}_i f_{ij}^0]^{-1} \tag{9.15}$$

This represents a generalisation of the balancing process for the usual journey-to-work model (Wilson, 1970). The parameters α, β, γ, ϕ, η and θ are determined *simultaneously* with the above balancing process via Newton-Raphson iteration.

For short time periods or in cases of a relative immobile population group, the numbers of movers M_i is a very small proportion ot the number of stayers M_{ii} in each region i. Especially if we omit the nesting constraint (9.9), the above formulation can be considerably simplified. In such cases, it can be shown that the summation $(\Sigma_{j\neq i} B_j^0 f_{ij}^0)$ is much less than unity, yielding the simple results

$$M_{ij} = M_j^0 (P_i^0 - U_i^0 - D_i^0) f_{ij}^0 / [\Sigma_{i\neq j} (P_i^0 - U_i^0 - D_i^0) f_{ij}^0] \tag{9.16}$$

In the parlance of Wilson (1970), this represents a simple destination-constrained model. Thus, during determination of the behavioural parameters, no concurrent balancing procedure is required.

Adjustment from Information Theory. Although the behavioural factors which we have chosen above seem plausible, there are further factors influencing migration which are not easy to identify via constraints. For example, some potential migrants just want a *change* from their habitual lifestyle. Furthermore, although we are modelling groups of individual migrants, many migration decisions are household based, with some members often making sacrifices for the sake of the household as a whole. Finally, housing prices and unemployment rates change over time, throwing up some doubts about the relevance of a comparative static analysis (see next Section). In addition, if housing prices express some sense of the *amenity* of a region, there may be leads or lags (more often the latter) in the provision of the infrastructure to confirm this amenity. Although these arguments point towards a dynamic approach, there is a clear way to take the influences of so-called *unexplained factors* into account, as well as allowing the influences of changes in identifiable factors to be evaluated. As discussed in Sect. 1.3.1, the hybrid entropy/information theory approach of Roy (1987) can be used, extending the pure information theory method of Snickars and Weibull (1977).

Let M_{ij}*, M_i* and M_{ii}* be the estimated base period model flows from (9.10) to (9.15). Also let $q_{ij} = M_{ij}^0/M_{ij}^*$, $q_i = M_i^0/M_i^*$ and $q_{ii} = M_{ii}^0/M_{ii}^*$ be the ratios of observed to modelled flows. Note that, if all these ratios are acceptably close to unity we need not proceed further - the regular entropy model is already a very good fit to the observations. Otherwise, q_{ij}, q_i and q_{ii} are normalised (just a cosmetic gesture) and then applied as prior probabilities to the original entropy objective in (9.1), yielding a transformed information theory objective I^0 as

$$I^0 = \Sigma_{ij} M_{ij} [\log (M_{ij}/q_{ij}) - 1] \qquad (9.17)$$

Now, if (9.17) is minimized with respect to M_{ij} {note the removal of the negative sign on (9.1) changes our maximization to minimization} under the identical set of constraints as applied to (9.1), we obtain a result with the important properties (i) the Lagrange multipliers of this modified problem are *identical* to those in the original entropy problem and (ii) the new modelled flows perfectly correspond to the observed flows M_{ij}^0. Upon setting $\bar{f}_{ij}^{\,0} = (q_{ij}/q_i) f_{ij}^{\,0}$, (9.10) becomes

$$M_{ij}/M_i = B_j^0 \bar{f}_{ij}^{\,0} / (\Sigma_{j \neq i} B_j^0 \bar{f}_{ij}^{\,0}) \equiv M_{ij}^0/M_i^0 \qquad (9.18)$$

and for the stayers and out-movers in (9.11) and (9.12)

$$M_{ii} = q_{ii} A_i^\theta \equiv M_{ii}^0 \quad ; \quad M_i = q_i A_i^\theta [\Sigma_{j \neq i} B_j^0 \bar{f}_{ij}^{\,0}]^\theta \equiv M_i^0 \qquad (9.19)$$

This is seen to be an information theory model which retains the explanatory factors contained within f_{ij} whilst perfectly reproducing the observed base period flows. Thus, it is likely to have improved predictive reliability over the pure entropy model, at least in the short term.

9.2.3 Use of Estimated Model for Projection

As discussed in Sect. 3.2.4, information is provided to the model estimation process via constraints, potentially enriching the model, reducing the uncertainty of the estimated probability distribution and yielding improved projection ability. Whilst the data for evaluation of the right-hand sides of the constraints is given for the base period, it becomes unknown for the projection time period if it reflects information *internal* to the structure of the model. For example, the base period in-migration totals M_j^0 into each region j used as right-hand sides of the associated constraints are required as *output* in the projection time period. As shown in Sect. 3.2.4, we adapt the formalism of Lesse (1982) to this problem by making invariant Legendre transforms of the original entropy Lagrangian defined as Z^0, that is the objective function (9.1) enhanced by addition of all the constraints multiplied by their respective Lagrange multipliers. Not only M_j^0, but also the right-hand sides of all the behavioural constraints, including that on the out-mover and stayer entropy, must emerge as output during the projection period. This implies that the transformed base period Lagrangian Z'^0 becomes as follows

$$Z^0 = Z^0 - \Sigma_j M_j^0 \, \partial Z^0/\partial M_j^0 - c^0 M^0 \, \partial Z^0/\partial c^0 M^0 - r^0 M^0 \, \partial Z^0/\partial r^0 M^0 - n^0 M^0$$
$$\partial Z^0/\partial n^0 M^0 - r^{l0} M^0 \, \partial Z^0/\partial r^{l0} M^0 - h^{r0} M^0 \, \partial Z^0/\partial h^{r0} M^0 - S^0 \, \partial Z^0/\partial S^0 \qquad (9.20)$$

This invariant transformation effectively removes the right-hand sides of the original Lagrangian, yielding the transformed version as

$$Z^0 = - \Sigma_{ij} M_{ij} [log(M_{ij}/q_{ij}) - 1 + \psi_j + \beta c_{ij}^{\,0} + \eta r_i^0 + \phi n_{ij}^{\,0} - \gamma(r_i^{\,0} - r_j^{\,0})$$
$$- \alpha(h_i^{\,0} - h_j^{\,0})] + \Sigma_i \lambda_i (P_i^{\,0} - U_i^{\,0} - D_i^{\,0} - M_{ii} - \Sigma_{j\neq i} M_{ij}) - (1-\theta)/\theta \{ \Sigma_i (\Sigma_{j\neq i} M_{ij})$$
$$[log (\Sigma_{j\neq i} M_{ij}) / q_i - 1] + M_{ii} [log (M_{ii}/q_{ii}) - 1] \} \qquad (9.21)$$

Now, for projection, the new explanatory variables are defined in (9.21) by removal of the zero superscripts. If this new Lagrangian is now maximised in terms of the flows, the following projection models emerge for the out-flows at a specified time period t

$$M_{ijt} / M_i^t = B_j^0 \, \bar{f}_{ij}^{\,t} \, / (\Sigma_{j\neq i} B_j^0 \, \bar{f}_{ij}^{\,t}) \qquad (9.22)$$

and for the stayers

$$M_{ii}^t = (P_i^t - U_i^t - D_i^t) / [\, 1 + (\Sigma_{j\neq i} B_j^0 \, \bar{f}_{ij}^{\,t})^\theta / q_i \,] \qquad (9.23)$$

and for the total out-movers

$$M_i^t = (P_i^t - U_i^t - D_i^t) / [\, 1 + q_i (\Sigma_{j\neq i} B_j^0 \, \bar{f}_{ij}^{\,t})^\theta \,] \qquad (9.24)$$

In this form, the above represents a special class or origin-constrained model, but with the behavioural expressions raised to the power of θ.

Now applying the demographic accounting principle, we can write the regional populations P_i^{t+1} at the next time period as

$$P_i^{t+1} = \Sigma_{j\neq i} M_{ji}^t + M_{ii}^t + I_i^t + B_i^t \qquad (9.25)$$

where I_i^t are the immigrants coming into region i from outside the system and B_i^t the births within region i during time period t. Such fundamental demographic data need to be provided from outside the model and can often be given alternative values within different scenarios.

As stressed in Sect. 3.2.4, the elegance of the above transformation procedure should not blind us to the fact that its success in projection depends partly on how stable over time are the base-period Lagrange multipliers, used as parameters in projection. As they can often be related to *elasticities*, similar questions arise for traditional economic models. We can check the stability of the parameters by estimating the model separately on the two or three most recent time periods for which data are available. If either the change is small, or we can identify a definite trend over time of the change, we can then select the values of the parameters to use for future time periods.

9.2.4 Empirical Performance

From Flood *et al.* (1991) and Roy and Flood (1992), some results are presented from a study on interregional migration in Australia. In general, the results of that study were disappointing with respect to model validation. Whereas the above model reduced the Root-Mean-Square error of the flows from between 35-65% for the regular gravity model to between 18-24% for our more sophisticated model, these errors are still unacceptably large. We did obtain some improvement by introducing an intrastate dummy variable to reflect some greater *à priori* chance of moving within one's own state. However, we are still left with the pecularities of the spatial distribution of population within Australia, whereby some intraregional moves (such as moves within Rest of Western Australia), lumped with the stayers in our model, involved longer distances than interregional moves in the more densely populated eastern states of Victoria and New South Wales. In future applications, we should at least try to separate out the stayers from the intraregional movers within the larger regions.

Another problem relates to equilibrium in the economy. Our 5-year data periods happened to span and partly submerge the effects of the 1982-84 recession. Thus, our attempt to model movement related to unemployment and rent differentials at 5 year intervals is quite shaky. However, these are the only intervals for which we have census data on interregional flows. The quarterly flow data are only available for interstate moves, and inevitably allocates intrastate moves to the stayers. This all points towards the necessity for developing a dynamic estimation procedure over shorter time periods, even though the available information is much less in inter-censal years. This is the objective of the following section.

9.3 A Two-Stage Dynamic Estimation Procedure

Although the above sequential comparative static procedure applied over several time periods may yield reasonable short-term projections in times of relatively stable economic conditions, it is incapable of identifying dynamic disequilibrium behaviour during strong economic upturns or downturns. In theory, the above model could be extended to be fully dynamic, estimating all the parameters *simultaneously* over a chosen set of previous time periods. However, model estimation would be a gargantuan effort. Thus, we would like to find an acceptable way to simplify the structure of the interaction model so that it can be estimated *simultaneously* over time.

In Weidlich and Haag (1988), an important contribution to this issue was made by the specification of a plausible interaction model in terms of generic utility, interaction and mobility components, with estimation of these components being performed using some alternative fitting techniques. In a second stage, lagged regression time series models in terms of characteristics of the origins, destinations and their interactions are fitted to these generic components (Reiner and Munz, 1990). Ledent (1986) suggested an alternative approach based on a gravity model framework, using the traditional power form of the impedance. Weidlich and Haag introduce an important symmetry assumption for the interaction terms, allowing endogenous determination of each of these factors. On the other hand, they implicitly impose a reciprocal relation between their *(i,j)* and *(j,i)* utility components, which causes any such component to relatively affect 'origin push' exactly inversely compared with 'destination pull'. By selectively combining the key insights of Weidlich, Haag and Ledent, as well as with the hybrid estimation procedure of Roy (1987), it is possible to overcome the above restrictions and to enhance the projection capability of the model. The admission of unsymmetrical interactions into the new approach allows intransitivities in the flow patterns to be accomodated, as discussed in Sect. 9.1.7. Then, time series interregional migration data can be used to estimate the origin push factors, the destination pull factors and the pairs of origin-destination interaction factors. Next, log-linear models are used to fit these factors in terms of selected demographic, economic and accessibility characteristics of the chosen set of regions, guided by the variables discovered to be significant by Flood *et al.* (1991). At the same time, a unified information theory approach integrates the model specification effect with the estimation step.

9.3.1 Relevant Previous Work

Weidlich and Haag Approach. The mean value equations developed in Chap. 3 of Weidlich and Haag (1988) are expressed in more conventional notation at any time period t as

$$M_{ij}^t = v_0^t P_i^t P_j^t \exp(u_j^t - u_i^t] f_{ij} \tag{9.26}$$

where M_{ij}^t is the number of moves between region i and region j during time period t, v_0^t is the global mobility (scaling) factor for time period t, P_i^t and P_j^t are the net populations in region i and j at the start of time period t, u_i^t is the generic 'utility' in region i at time period t and f_{ij} is the interaction factor between regions i and j, assuming that $f_{ij} = f_{ji}$. Note that, the pair-wise interaction factors are taken to be time-independent.

In order to determine the unknown mobility, utilities and interaction factors, two fitting procedures are suggested. In the first, one minimizes simultaneously over all time periods and all regional pairs *(i,j)* the sum of the squares of the differences between the log of the observed flows $M_{ij}^{0\,t}$ and the log of the modelled flows M_{ij}^t as expressed via (9.26). Alternatively, one can express the squared differences in terms of the flows themselves, rather than the log of the flows, yielding a non-linear procedure. Note that, neither of these approaches corresponds with the well-proven method of maximum likelihood.

With the minimization procedure yielding $[n\,(n+1)/2]$ relations for the $[n\,(n-1)]$ off-diagonal flows M_{ij}^t (omitting intraregional moves, as before), the method will clearly usually give a very good fit to observations. This has been verified in applications of the model. On the other hand, the occurrence of the utility terms as $exp\ (u_j^t - u_i^t)$ in equation (9.26) for the flows M_{ij}^t implies a special type of asymmetry in the influence of the utility terms on opposing flows between regional pairs, which may not correspond well with observations. In other words, the relative incentive, for instance, to move from the higher to the lower unemployment region of a pair is the inverse of the disincentive to move in the opposite direction. The authors did experiment with a weighted utility difference $exp\ (u_j^t - \mu\, u_i^t)$, indicating that μ came out to be close to unity. However, it would have been more general to have made coefficient μ region-specific. Such flexibility is implicit in the approach of Ledent (1986). However, Ledent did not introduce the important symmetry assumption $f_{ij} = f_{ji}$ of the Weidlich and Haag model. which (i) yields endogenous values of the 'connectivity' f_{ij} between each regional pair and (ii) gives an excellent fit to the observed flows.

After the generic utilities, interaction and mobility factors have been evaluated, the second step is to relate them to the economic and demographic characteristics of the regions using a lagged time series analysis. A special optimally-lagged procedure was introduced by Reiner and Munz (1990) from related work in laser theory. An important future task is to compare this approach with other time series methods used in spatial statistics.

Ledent Approach. The first step of the method of Ledent (1986) is the base period calibration of a doubly-constrained gravity model of the form

$$M_{ij} = k\, v_i\, w_j\, d_{ij}^{-h} \tag{9.27}$$

where k is a scaling factor, v_i is the origin push factor of region i, w_j is the destination pull factor of region j, d_{ij} is the average road distance between regions i and j and h is the gravity index. In the calibration v_i is calibrated to fit the observed out-migration totals from region i, w_j to fit the observed in-migration totals into region j and h to fit the total travel distance $(\Sigma_{ij}\, M_{ij}^0\, d_{ij})$.

In comparing (9.26) with (9.27), it is noted that the latter does not impose the restrictive symmetry conditions in terms of utility differences, allowing v_i and w_j to be evaluated independently. However, its characterisation of the regional interaction effects just through the one parameter h is a disadvantage.

In the second step of the method, the origin and destination factors v_i and w_j are explained in terms of attributes of the origin and destination regions respectively, using log-linear regression. In contrast to the Weidlich and Haag approach, Ledent does not appear to introduce any lags in movement in response to economic change and the time series estimation is performed sequentially instead of simultaneously. Thus, it shares a lot of the properties of a comparative static model, similar to the model formulated in the earlier part of this chapter, but without some of its endogenous features. We are left, as usual, with a multitude of advantages and disadvantages of former models in our search for a tractable fully dynamic framework.

9.3.2 New Model Specification and Estimation

With the advantage of hindsight, it is possible to combine the virtues of the Weidlich and Haag approach with those of the Ledent procedure within the information-theoretic framework of Roy (1987). Initially, plausible migration relationships will be introduced and the relevant parameters estimated. Finally, an analogous result is produced more formally from information theory.

A Plausible Formulation - First Step. A feature of the estimation of discrete choice models (Hensher and Johnson, 1981) is the strict separation between pure linear size effects and economic/behavioural effects. This property seems particularly relevant in our current context, allowing the linear population totals P_i^t and P_j^t to be separated out in the first step (the generic model), with the lagged time series analysis in the second step concentrating on economic effects and the more secondary non-linear population *saturation* effects. Thus, a plausible generic model can be expressed as

$$M_{ij}^t = v_0^t P_i^t P_j^t A_i^t B_j^t f_{ij} \tag{9.28}$$

where v_0^t is the overall mobility factor [see (9.26)], A_i^t is the origin push factor from region i, B_j^t is the destination pull factor to region j and f_{ij} is the symmetrical interaction factor introduced by Weidlich and Haag. The estimated model must reproduce the total number of movers M^{ot} via the constraint

$$\Sigma_{ij} M_{ij}^t = M^{0t} \tag{9.29}$$

In order to estimate the parameters v_0^t, A_i^t, B_j^t and f_{ij}, a maximum likelihood approach is adopted in the form of minimization of the objective Z

$$Z = \Sigma_{t=1}^{T} \Sigma_{ij} M_{ij}^{0t} \log M_{ij}^t \tag{9.30}$$

where the modelled flows M_{ij}^t are given from (9.28) and M_{ij}^{0t} are the movement observations at base period t within the estimation time periods 1 to T. With differentiation of (9.30) with respect to v_0^t, relation (9.29) emerges. Upon differentiation with respect to the origin push factors A_i^t and the destination pull factors B_j^t, the following equations emerge

$$\begin{aligned} A_i^t &= M_i^{ot} [v_0^t P_i^t \Sigma_j P_j^t B_j^t f_{ij}]^{-1} \\ B_j^t &= M_j^{0t} [v_0^t P_j^t \Sigma_i P_i^t A_i^t f_{ij}]^{-1} \end{aligned} \tag{9.31}$$

In the above, M_i^{0t} is the observed number of out-movers from region i and M_j^{0t} the observed number of in-movers into region j in estimation period t. Note that, by substitution from (9.28), relations (9.31) can be seen to reduce to the following classical origin and destination constraints

$$\Sigma_j M_{ij}^t = M_i^{0t} \quad ; \quad \Sigma_i M_{ij}^t = M_j^{0t} \tag{9.32}$$

The symmetrical interaction factors f_{ij} are also obtained from non-linear relations which reduce to

$$\Sigma_t [M_{ij}^t + M_{ji}^t] = \Sigma_t [M_{ij}^{0t} + M_{ji}^{0t}] \tag{9.33}$$

These are quite a strong set of constraints, implying that, for *each (i,j)* regional pair, the sum of the opposing modelled flows summed over all the estimation time periods must equal the corresponding observed flow sums. The satisfaction of these constraints normally ensures that the model yields a good fit to observations.

The system of equations (9.29) and (9.33) [upon substitution of (9.28)] and (9.31) can be solved iteratively for the A_i^t, B_j^t, v_0^t and f_{ij} values as a generalisation of the usual bi-proportional method. Although (9.32) implies that (9.29) will be satisfied automatically, relation (9.29) should be solved explicitly for v_0^t, as its evolution over time is a useful measure of global mobility of the population as a whole. As the system is now twice over-determined, two factors are set arbitrarily, such as $A_1^t = 1$ and $B_1^t = 1$, with v_0^t and the rest of A_i^t and B_j^t being automatically scaled accordingly.

Note that, if the Weidlich and Haag model in (9..26) were estimated by the maximum likelihood criterion (9.30), the relations for the utilities u_i can be shown to reduce to the following conditions

$$\Sigma_j M_{ij}^t - \Sigma_j M_{ij}^t = \Sigma_j M_{ji}^{0t} - \Sigma_j M_{ij}^{0t} \tag{9.34}$$

This set of constraints is seen to cause the model to merely reproduce the *net* migration totals in each region i. This is rather odd for a model of *gross* migration. On the other hand, we have in (9.32) two sets of *gross* migration constraints, one on the total out-flows and the other on the total in-flows. Despite this, with its large number of equations for determining the f_{ij} factors, the Weidlich and Haag method does usually provide a good fit to observations. At the same time, the restrictive nature of the constraints (9.34) is unnecessary, as shown in the new formulation above.

A More Formal Derivation. From the constraints which have emerged above as our maximum likelihood equations, it is straightforward to obtain an identical result using an information theory formulation expressed as

$$I = \Sigma_{t=1}^T \Sigma_{ij} M_{ij}^t [\log (M_{ij}^t / P_i^t P_j^t) - 1] \tag{9.35}$$

If (9.35) is minimised under constraints (9.29), (9.32) and (9.33), we perfectly reproduce not only the model form (9.28), but also identical values of the model parameters. This is not surprising, especially due to the intimate relations between entropy-type models and maximum likelihood approaches, as demonstrated by Karlqvist and Marksjö (1971).

The new formulation in (9.28), with its symmetrical interaction terms $f_{ij} = f_{ji}$ is not capable of reproducing intransitive flows, which were identified by Goodchild and Smith (1980). However, this deficiency can be overcome by using the hybrid entropy/information theory technique of Roy (1987). If the flow results for the estimated model (9.28) are denoted as $\tilde{M}_{ij}^t$, the application of prior probabilities $q_{ij}^t = M_{ij}^{0t} / \tilde{M}_{ij}^t$ to the denominator of the log terms in (9.35) will result in a model which (i) perfectly reproduces the observed flows M_{ij}^{0t} and (ii) yields identical parameters to those in the original (9.32). The revised model now becomes

$$M_{ij}^t = v_0^t P_i^t P_j^t A_i^t B_j^t \tilde{f}_{ij}^t \tag{9.36}$$

where $\tilde{f}_{ij}^t = q_{ij}^t f_{ij}$. As there is no reason for q_{ij}^t to equal q_{ji}^t, the enhanced model (9.36) can pick up intransitive flows.

Model Formulation - Second Step. Having determined the generic origin factors A_i^t, the destination factors B_j^t and the interaction factors f_{ij}, it is now time to use one or more variants of regression analysis to explain them in terms of distance, population and behavioural characteristics of the regions. In the underlying entropy or logit-type models, these characteristics are always present in exponential form; hence the log of the factors is fitted in terms of an additive lagged utility function Z of the behavioural variables enhanced by a logarithmic saturation effect. For example, for the origin push factors A_i^t, we minimize the function Z_A expressed as

$$Z_A = \Sigma_{t=1}^T \Sigma_i [\, \log A_i^t - \{a + \alpha \log P_i^t + \Sigma_k \eta_k u_{ik}(t-\tau)\} \,]^2 \tag{9.37}$$

In this objective function *'a'* is a shift parameter, $\alpha > 0$ represents a population saturation effect and η_k is the parameter on the kth behavioural variable $u_{ik}(t-\tau)$, taken to be lagged by τ estimation time periods. As an alternative to the above standard regression procedure, one may *simultaneously* determine optimal lags via the approach of Rainer and Munz (1990), which is based on certain approaches from laser theory.

9.3.3 Simple Application of the Two-Stage Model

The 1986 Australian Census data permit internal gross migration movements during 1985-86 to be disaggregated by region as well as by age cohort. In this initial trial of the new approach, 22 regions and 4 age cohorts are used, as in Flood *et al.* (1991). With just the single year of such data, it was clearly not possible to carry out a time series analysis, nor to introduce the lag τ. As shown above, the first step is to evaluate the generic factors within (9.28) and the following equations, checking the plausibility of the results, Finally, a log-linear analysis, as in (9.37), is performed to identify the behavioural variables which yield the best fit to the generic factors.

Estimation of Generic Factors - First Stage. Using an enhancement of the standard iterative balancing procedure, equations (9.29), (9.31) and (9.33) are solved for each of the 4 age cohorts. The results for the global mobility parameter v_0 are shown below:

Age Group	0-14	15-29	30-54	55 and over
μ_0	0.591	0.965	0.292	0.384

These results confirm the expected high mobility 15-29 age group, as well as the low mobility of the more family-oriented 30-54 group. For the oldest group, there is an increase of mobility over the 30-54 group, reflecting the more open location options available to retirees. In interpretation of these results, we note that intraregional movers have not been included in the analysis, being merged into the group of stayers. This means, in particular, that important intra-metropolitan moves have been neglected, which are especially important for couples in the process of starting a family.

Rather than tabulating all the 176 balancing factors A_i and B_j, we just mention certain key characteristics of the results. First, regions with both high origin push and high destination pull factors generally have high levels of gross migration, but not necessarily high net migration levels (eg. the tropical 'frontier' region of Darwin). Regions with relatively low levels of destination pull will generally be net losers and vice versa (eg. older metropolitan regions vs sunbelt regions). In fact, the highest destination pull factor occurs for the 55 and over group to Australia's major sunbelt destination, the Gold Coast. On the other hand, for the 15-29 age group, the frontier region of Darwin had the highest destination pull factor. The two old metropolitan regions of Melbourne and Sydney have the distinction of possessing the lowest destination pull factors for all 4 groups. At the same time, Melbourne and Sydney have among the lowest origin push factors, reflecting an amenity value for those already located there. With the results being normalised by extraction of the origin and destination populations, these push and pull factors are meaningful when compared relatively. The lowest origin push factor occurs for the 55 and over age group from the Adelaide region. Adelaide is a relatively industrial city and its retirees may well have less mobility than the average. Of course, the balancing factors represent the accessibility-independent component of migration propensity, with further multiplication by the accessibility factors f_{ij} being required to evaluate the modelled flows.

For those of us with local knowledge, the results seem to be intuitively plausible. Also, RMS errors were about 15% for the 0-14, 30-54 and 55 and over age groups, but only about 7.5% for the highly mobile 15-29 group. While the latter result is quite acceptable, the 15% error cases have probably been influenced by the variable nature of the regions. With almost half of Australia's population being located within a 100 kilometer radius of Melbourne and Sydney, many of the remaining regions are large and sparse. For instance, whereas movers within the extremely large region Rest of Western Australia are omitted, movers between adjacent smaller regions (eg. the nearby northern cities of Townsville and Cairns) are included. This problem is inherent to the geography and unusual population distribution of Australia, and cannot substantially be overcome.

Regression Model - Second Stage. Multiple step-wise regressions were performed for the set of origin push factors $log\ A_i$ and again for the destination pull factors $log\ B_j$ for each of the four age groups. The explanatory variables include: Average Income; Unemployment Rate; Average Housing Rent; Size of Workforce; Average Housing Rent (excluding lowest public category); Number of Caravans (Trailers in the US); Number of Hotels/Motels; Warm Climate Dummy.

In summary, the R^2 values varied between 0.75 and 0.85, with the best value occurring for the origin push of the mobile 15-29 group. Now, despite the preliminary extraction of linear population influences from the push and pull factors, the population coefficient remains the largest and most significant influence in the model in all cases. Also, the coefficient is always negative, varying between -0.21 to -0.38. Thus, this yields the net effect of population on flows as lying between $P^{0.62}$ to $P^{0.79}$. For large regions, this implies a population saturation effect on destination pull and a population retention effect on origin push. These values are clearly influenced by the low gross migration levels in and out of the two major regions of Sydney and Melbourne.

In examining the behavioural influences, we start with the 0-14 age group. The number of caravans (an environmental amenity variable) was the most significant behavioural variable and that with the largest standardised coefficient for origin push. This indicates a tendency for the associated family groups to move out of coastal resort regions which do not contain capital cities. On the other hand, for destination pull, both high rent and high unemployment rate are the major attractors. With higher rents in urban areas and higher unemployment in rural areas, this result is ambiguous.

For the 15-29 group, the origin push is primarily explained by movement out of high unemployment regions, whereas destination pull is explained mainly by movement into high rent regions, with associated higher amenities.

The 30-54 group, as with the 0-14 group which accompany them, again have number of caravans as their most significant and largest behavioural influence on origin push. Income and unemployment are also fairly significant. However, for destination pull, high unemployment turns out to be the major attractor! This reflects the negative net migration levels in the low unemployment regions of Melbourne and Sydney. The larger net migration gainers, the more 'peripheral' state of Queensland and Perth in Western Australia, had higher unemployment than the major centres of Melbourne and Sydney.

Finally, we examine the results for the 55 and over group. For the first time, the regression gave spurious results here, that is, for origin push. This problem requires further examination. On the other hand, for destination pull, the unemployment variable was both large and highly significant (0.02% level). This confirms the movement of many in this age group to resort and rural areas, which traditionally have much higher unemployment levels than in the capital cities. Of course, the rents are considerably lower, which is a key consideration for the many retirees who still rely on welfare.

An attempt was also made to fit the interaction coefficients f_{ij} to either a negative power form or a negative exponential function of distance. However, the best R^2 result of 0.34 does not support the use of either of these traditional measures of spatial deterrence. This confirms the results in Flood *et al.* (1991), where an intrastate movement dummy variable gave much better results than a negative exponential function of distance. This reflects a preference to make moves within one's own state, even if one is relatively close to the border of another state. The results are also likely to be influenced by some extremely large differences in the sizes of the 22 regions.

In the overall evaluation of these regression results with respect to model structure, several points should be made:

(i) data on economic variables, such as unemployment, are usually only available for the regional population as a whole, rather than for the movers. For regions and age groups which show low mobility, these values may be quite unrepresentative.

(ii) certain other variables, such as rents, are often only available for the eight States and Territories or for the capital cities. They may not be obtainable at the regional level.

(iii) with the availability of data being even worse in intercensal years, it will be difficult to attempt with confidence the ultimate aim of the work, a lagged time series analysis.

In spite of the above limitations, the suggested approach seems both plausible and efficient. The two-stage approach, by allowing a small set of generic factors to be estimated first, postponing the introduction and identification of the behavioural factors until the second stage, reduces the computational effort for a single time period compared with the comparative static approach. These savings allow, in principle, a lagged time series estimation to be accomplished.

9.3.4 Future Issues

We still face the major problem of the paucity of regional data in the intercensal years. This is especially relevant in our professed aim of making a detailed time series estimation, with time periods being typically one year rather than five years. In Australia, we now have quarterly gross migration data for interstate moves. It turns out that information theory is an ideal instrument for handling data at different levels of spatial aggregation within a unified analysis. In other words, we can endeavour to fit an interregional model over one year intervals to the interregional data available each 5 years at the Census and the interstate data available in each of the intervening years. In this analysis, we need to express the interstate moves as summations of the underlying interregional moves, with the model then evaluating the latter, even in the intercensal years. It is difficult to say in advance the improvements which are achievable over a model just estimated over a set of 5-year periods. At least, with one year intervals, there is less chance of skipping over a major economic event, such as a recession.

A further issue, touched on earlier, is that regional economic data is available for the region as a whole, rather than for the subset of people who move to or from that region within a certain time period. The only way to gauge the possible importance of this problem is to make special surveys of movers. Initially, this could be a limited survey to see if the problem is serious. If so, support should be then sought to make more comprehensive surveys.

Future Challenges

An approach has been described for not only transforming certain models to be probabilistic rather than deterministic, but also to provide a high level of flexibility in relation to the use of data in the estimation time period compared to that in the projection time period. Although other model frameworks are available, the entropy approach has the advantage of integrating a microlevel in terms of individual economic agents with a macrolevel in terms of the market segments and/or spatial zones or regions for which data is commonly available. At the same time, it is clear that skills in defining marker segmentation, both at the spatial and behavioural levels, need to be developed in cases where we have data to support alternative forms of market segmentation. In this way, we may succeed in reducing within-segment variance of the characteristics of the market, allowing most of it to be expressed between segments. This is the aim for all spatial models, where contiguity inevitably constrains segmentation.

Scholars with a background in spatial econometrics who are examining this work may be concerned at the paucity of error analysis (see end of Chap. 1). Although several goodness-of-fit statistics are potentially available upon model estimation, there is just one measure of robustness and one measure of event correlation. As with the method of maximum likelihood in the past, scholars with experience in spatial statistics may well be able to enrich the error structure of this class of model. This is particularly pertinent because of the correspondence between maximum entropy and maximum likelihood solutions, established by Karlqvist and Marksjö (1971). In the same spirit, the links between spatial interaction and spatial autocorrelation, developed by Getis (1991), should be further investigated. Similarly, the links with neural networks established by Thill and Mozolin (2000) should be further investigated. At the same time, it is certainly comforting that Smith (1990) demonstrated that for sufficiently large populations, the most probable state is *overwhelmingly* the most probable. In a recent private communication with the author, Smith stated that "*Basically MPS (Most Probable State) analysis can be viewed as an elaborate application of the Law of Large Numbers, with special emphasis on the sharper exponential version of this law, known as "Large-Deviation Theory". But while these exponential bounds are very appealing, they still offer few guidelines for practical applications. What seems to be required here is a systematic investigation of MPS in terms of stochastic simulations. I have done some recent work with Markov Chain Monte Carlo (MCMC) simulations [that are currently revolutionizing the practice of Bayesian statistical analysis]. I believe that these methods offer a potential way to gain insights into the practical consequences of the MPS approach. For example, from a planner's perspective, it does little good to know what the "overwhelmingly most probable" response to a given policy will be unless this can be translated into actual confidence bounds on key response variables.*" It was most unfortunate that the Regional Science Department at University of Pennsylvania folded, taking Tony Smith away from making the further progress sketched above. In any case, it's likely that some areas of application of entropy type models will yield adequate confidence bounds and others not. For example, many results from these models are observed to have the comforting property that the relative error in the significant flows is smaller that that in the minor flows. Also, our entropy solutions represent an interpolation, bounded from above by the pure quantity-based attractor approach implied by the trade model of Ohlin (1933) and from below by classical deterministic results based, for example, on the theory of the firm (Intriligator, 1971).

One major emphasis is a clear delineation between estimation and projection forms of the models. This was not yet clearly spelled out in the seminal Wilson (1967, 1970) works. In fact many observers inferred, for example, that average travel costs were either constant or needed to be input as *exogenous* data at projection time periods. As such quantities are clearly desirable model *outputs*, such interpretations appeared to reduce the potential value of the models for policy purposes. Of course, the formulations introduced above, where certain estimated Lagrange multipliers are transformed into parameters for projection, does not fully dispose of the problem. As with elasticities, trends for change of these parameters need to be identified via re-estimation of the models at different time periods.

Although the enhancement of some of the key deterministic models of microeconomics into various probabilistic forms may well be an advantage in using the models for projection, model assumptions are often more critical. This concern is reflected in this book by considerable attention to imperfect competition. In addition, issues such as increasing returns and spatial agglomeration effects need to be understood further and accounted for in the models. This book has mainly concentrated on short run demand/supply modelling, with attention to long run location modelling confined to housing, retail facilities and knowledge firms. A major challenge is to assess the implications of the recent fundamental work of Fujita and Thisse (2002) on all the locational models that have been presented here. This includes a wider appreciation of nested dynamic approaches, confined to interregional migration in this book.

References

Citations

Alonso, W. (1964): Location and Land Use. Harvard University Press, Cambridge

Anas, A. (1983): Discrete Choice Theory, Information Theory and the Multinomial Logit and Gravity Models. Transportation Research B 17, 13-23

Angel, S., Hyman, G.M. (1970): Urban Velocity Fields. Environment and Planning 2, 211-224

Angel, S., Hyman, G.M. (1972): Urban Spatial Interaction. Environment and Planning, 4, 99-118

Armington, P.S. (1969): A Theory of Demand for Products Distinguished by Place of Production. IMF Staff Papers, 159-176

Ashley, B. (1997): Are Malls in America's Future?. Arthur Andersen Retailing Issues Letter, 9(6), Texas A&M University Center for Retailing Studies

Batten, D.F. (1983): Spatial Analysis of Interacting Economies. Kluwer-Nijhoff, Boston

Baxter, M.J. (1985): Quasi-Likelihood Estimation and Diagnostic Statistics for Spatial Interaction Models. Environment and Planning A 17, 1627-1635

Beckmann, M.J., Thisse, J.-F. (1986): The Location of Production Activities. In: Nijkamp, P. (Ed.): Handbook of Regional and Urban Economics 1, Regional Economics. North-Holland, Amsterdam, 21-95

Beckmann, M.J. (1994): On Knowledge Networks in Science: Collaboration Among Equals. Annals of Regional Science 28, 233-242

Bhat, C.R., Koppelman, F.S. (1993): A Conceptual Framework of Individual Activity Program Generation. Transportation Research, 27A(6), 433-436

Binger, B.R. (1988): Microeconomics with Calculus. Harper Collins, United States

Bonsall, P. (1975): Approaches to the Prediction of Intra-Zonal Interactions. Working Paper 65, Institute for Transport Studies, University of Leeds

Bowman, J., Bradley, M., Shiftan, Y., Lawton, K., Ben-Akiva, M. (1998): Demonstration of an Activity Based Model System for Portland. Proceedings of the 8th World Conference on Transport Research, Antwerp, Belgium

Boyce, D.E., Chon, K.S., Lee, Y.J., Lin, K.T., Le Blanc, L.J. (1983): Implementation and Computational Issues for Combined Models of Location, Destination, Mode and Route Choice. Environment and Planning A 15, 1`219-1230

Bröcker, J. (1989): How To Eliminate Certain Defects of the Potential Formula. Environment and Planning A 21, 817-830

Cambridge University, Department of Applied Economics (1963): Input-Output Relationships, 1954-1966. In: A Programme for Growth 3, Chapman and Hall, London

Cesario, F.J. (1973): A Note on the Entropy Model of Trip Distribution. Transportation Research 27, 331-333

Champernowne, A.F., Williams, H.W.C.L., Coelho, J.D. (1976): Some Comments on Urban Travel Demand Analysis and Economic Evaluation of Travel Plans. Journal of Transport Economics and Policy 10, 276-285

Chenery, H. (1953): Regional Analysis, In: Chenery, H., Clark, P., Pinna, V.(Eds.): The Structure and Growth of the Italian Economy. US Mutual Security Agency, Rome

Chenery, H.B., Clark, P. (1959): Interindustry Economics. Wiley, New York

Dellaert, B., Arentze, E., Bierlaire, M., Borgers A., Timmermans, H. (1998): Investigating Consumers' Tendency to Combine Multiple Shopping Purposes and Destinations. Journal of Marketing Research, 35(2), 177-188

Dixit, A.K., Stiglitz, J.E. (1977): Monopolistic Competition and Optimum Product Diversity. American Economic Review 67, 297-308

Dorigo, G., Tobler, W.R. (1983): Push-Pull Migration Laws. Annals of the Association of American Geographers 73, 1-17

Eaton, B.C., Lipsey, R.G. (1979): Comparison Shopping and Clustering of Homogeneous Firms. Journal of Regional Science 19, 421-435

Erlander, S. (1980): Optimal Spatial Interaction and the Gravity Model. Lecture Notes in Economics and Mathematical Systems 173, Springer, Berlin

Evans, S.P. (1973): A Relationship Between the Gravity Model for Trip Distribution and the Transportation Problem in Linear Programming. Tansportation Research 7, 39-61

Fisk, C., Brown, G.R. (1975): A Note on the Entropy Formulation of Distribution Models. Operational Research Quarterly 26, 755-758

Flood, J., Maher, C.J., Newton, P.W., Roy, J.R. (1991): Dterminants of Internal Migration in Australia. Final Report, Department of Industry, Technology and Commerce, Canberra

Fotheringham, A.S. (1983a): A New Set of Spatial Interaction Models: The Theory of Competing Destinations. Environment and Planning A 15, 15-36

Fotheringham A.S. (1983b): Some Theoretical Aspects of Destination Choice and Their Relevance to Production-Constrained Gravity Models. Environment and Planning A 15, 1121-1132

Fotheringham, A.S., Dignan, T. (1984): Further Contributions to a General Theory of Movement. Annals of the Association of American Geographers 74, 620-633

Fotheringham, A.S. (1986): Modelling Hierarchical Destination Choice. Environment and Planning A 18, 401-418

Fotheringham, A.S., Knudsen, D.C. (1986): Modeling Discontinuous Change in Retailing Systems: Extensions of the Harris-Wilson Framework with Results from a Simulated Urban Retailing System. Geographical Analysis 18, 295-312

Fratar, T. (1954): Vehicular Trip Distribution by Successive Approximation. Traffic Quarterly 8, 53-65

Frost, M.E., Spence, M.A. (1995): The Rediscovery of Accessibility and Economic Potential: The Critical Issue of Self-Potential. Environment and Planning A 27, 1833-1848

Fujita, M. (1989): Urban Economic Theory: Land Use and City Size. Cambridge University Press, Cambridge.

Fujita, M., Krugman, P.R., Venables, A.J. (1999): The Spatial Economy. MIT Press, Cambridge, Mass.

Getis A. (1991): Spatial Interaction and Spatial Autocorrelation: A Cross-Product Approach. Environment and Planning A 23, 1269-1277

Gonçalves, M.B., Ulysséa-Neto, I. (1993): The Development of a New Gravity-Opportunity Model for Trip Distribution. Environment and Planning A 25, 817-826

Goodchild, M.F., Smith, T.R. (1980): Intransitivity, the Spatial Interaction Model and U.S. Migration Streams. Environment and Planning A 12, 1131-1144

Hägerstrand, T. (1970): What About People in Regional Science?. Papers of the Regional Science Association 24, 7-21

Harris, B., Wilson, A.G. (1978): Equilibrium Values and Dynamics of Attractiveness Terms in Production-Constrained Spatial Interaction Models. Environment and Planning A 10, 371-388

Hensher, D.A., Johnson, L.W. (1980): Applied Discrete-Choice Modelling. Croom Helm, London, Wiley, New York

Herbert, J., Stevens, B. (1960): A Model for the Distribution of Residential Activity in Urban Areas. Journal of Regional Science 2, 21-36

Hewings, G.J.D. (1985): Regional Input-Output Analysis. Beverly Hills, Sage Publications Inc.

Holmberg, I. (1990): Models of Population Dynamics - An Information Theory Approach. CERUM Working Paper 1990:8, University of Umeå

Hotelling, H. (1929): Stability in Competition. Economic Journal 39, 41-57

Hotelling, H. (1932): Edgeworth's Taxation Paradox and the Nature of Demand and Supply Functions. Journal of Political Economy 40, 577-616

Huff, D. (1964): Defining and Estimating a Trading Area. Journal of Marketing 28, 34-38

Intriligator, M..D. (1971): Mathematical Optimization and Economic Theory. Prentice-Hall, Englewood Cliffs, N.J.

Isard, W. et.al. (1998): Methods of Interregional and Regional Analysis. Ashgate

Johansson, B. (1991): Regional Industrial Analysis and Vintage Dynamics. Annals of Regional Science, 23, 1-18

Jones, P..M. (1976): The Analysis and Modelling of Multi-Trip and Multi-Purpose Journeys. Working Paper 6, Oxford University Transport Studies Unit

Jones, P.M., Dix, M.C., Clarke, M.I., Heggie, I.G. (1983): Understanding Travel Behaviour. Gower Press, Aldershot, U.K.

Karlqvist, A., Marksjö, B. (1971): Statistical Urban Models. Environment and Planning 3, 83-98

Kitamura, R. (1984): Incorporating Trip Chaining into Analysis of Destination Choice. Transportation Research B 18, 67-81

Kobayashi, K., Batten, D.F., Andersson, A. (1991): The Sequential Location of Knowledge Oriented Firms over Time and Space. Papers in Regional Science 70, 381-397

Kobayashi, K., Roy, J.R., Fukuyama, K. (1998): Contacts with Agreements: Towards Face-to-Face Communication Modelling. Annals of Regional Science 32, 389-406

Kolstad, C.D., Mathiesen, L. (1991): Computing Equilibria in Imperfectly Competitive Markets. In: Güvenen, O., Labys, W., Lesourd, J.B. (Eds.): International Commodity Market Models, Chapman and Hall, London, 51-70

Koopmans, T.C. (1949): Optimum Utilisation of the Transportation System. Econometrica 17, Suppl.: 136-146

Kullback, S. (1959): Information Theory and Statistics. Wiley, New York

Lakshmanan, T.R., Hansen, W.G. (1966): A Retail Market Potential Model. Journal of the American Institute of Planners 31, 134-143

Launhardt, W. (1885): Mathematische Begründung der Volkswirtschaftslehre. Teubner, B.G., Leipzig (English Translation: (1993): Mathematical Principles of Economics. Edward Elgar, Aldershot)

Ledent, J. (1987): Forecasting Interregional Migration: An Economic-Demographic Approach. In: Isserman, A.M. (Ed.): Population Change and the Economy: Social Science Theories and Models, Kluwer-Nijhoff, Boston

Leontief, W.W. (1951): The Structure of the American Economy, 1919-1939. Oxford University Press, New York

Leontief, W.W. et al. (1953): Studies in the Structure of the American Economy. Oxford University Press, New York

Leontief, W. and Strout, A. (1963): Multi-Regional Input-Output Analysis. In: Barna, T. (Ed.): Structural Interdependence and Economic Development. London, Macmillan

Lerman, S.R. (1979): The Use of Disaggregate Choice Models in Semi-Markov Process Models of Trip Chaining Behaviour. Transportation Science 13, 273-291

Lesse, P.F., Brotchie, J.F., Roy, J.R., Sharpe, R. (1978): A New Philosophy for Regional Modelling. Proceedings of the 3rd Conference of the Regional Science Association, Melbourne

Lesse, P.F. (1982): A Phenomenological Theory of Socio-Economic Systems with Spatial Interactions. Environment and Planning A 14, 869-888

Martínez, F.J. (1992): The Bid-Choice Land Use Model: An Integrated Economic Framework. Environment and Planning A 24, 871-885

Martínez, F.J., Donoso, P. (1995): MUSSA Model : The Theoretical Framework. Modelling Transport Systems. In: Hensher, D., King, J., Oum, T. (Eds.): Proceedings 7th World Conference on Transportation Research WCTR, 2, Pergamon, 333-343

Mazurkiewicz, L. (1985): A Statistical Model of a Multitrip Spatial-Interaction Pattern. Environment and Planning A 17, 1533-1539

McAffe, P. and McMillan, J. (1987): Auctions and Bidding, J. Econ. Liter., XXV, 699-738

McFadden, D. (1981): Econometric Models of Population Choice. In: Manski, G.F., McFadden, D. (Eds.): Structural Analysis of Discrete Data with Econometric Applications, MIT Press, Cambridge, Mass.

Mills, E.S., Hamilton, B.W. (1972): Urban Economics. Scott, Foresman and Company, Glenview, Illinois

Miyagi, T. (1991): A Computational Procedure for Determination of Oligopolistic Spatial Price Equilibrium. Papers in Regional Science, 70, 185-200

Mohring, H. (1993): Land Rents and Transport Improvements: Some Urban Parables. Transportation 20, 267-283

Mueser, P.R., White, M.J. (1989): Explaining the Association Between Rates of In-Migration and Out-Migration. Papers of the Regional Science Association, 67, 121-134

Mun, Se-il, Yoshikawa, K. (1993): Communication Among Firms, Traffic Congestion and Office Agglomeration. Annals of Regional Science 27, 61-77

Mun, Se-il, Hutchinson, B.G. (1995): Empirical Analysis of Office Rent and Agglomeration Economy: A Case of Toronto. Journal of Regional Science, 35, 3, 437-455.

Munroe, D.K., Hewings, G.J.D. (1999): The Role of Intraindustry Trade in Interregional Trade in the Midwest of the US. Discussion Paper, 99-T-7, Regional Economics Applications Laboratory, University of Illinois, Urbana.

Nagurney, A. (1993): Network Economics. Kluwer Academic Publishers, Dordrecht, Boston, London

Nagurney, A., Dupuis, P., Zhang, D. (1994): A Dynamical Systems Approach for Network Oligopolies and Variational Inequalities. Annals of Regional Science, 28, 263-283

Odland, J. (1988): Sources of Change in the Process of Population Redistribution in the United States. Environment and Planning A 20, 789-809

Ohlin, B.G. (1933): Interregional and International Trade. Harvard University Press, Cambridge, Mass.

O,Kelly, M.E. (1981): A Model for the Demand for Retail Facilities, Incorporating Multistop, Multipurpose Trips. Geographical Analysis 13, 134-148

Plane, D.A. (1982): An Information Theoretic Approach to the Estimation of Migration Flows. Journal of Regional Science 22, 441-456

Plane, D.A., Rogerson, P.A. (1985): Economic-Demographic Models for Forecasting Interregional Migration. Environment and Planning A 17, 185-198

Plane, D.A. (1987): The Geographic Components of Change in a Migration System. Geographical Analysis 19, 283-299

Polenske, K.R. (1980): The U.S Multiregional Input-Output Accounts and Model. Lexington Books (D.C. Heath and Co.), Lexington, Mass.

Puu, T. (1979): The Allocation of Road Capital in Two-Dimensional Space. North-Holland, Amsterdam

Reichman S. (1976): Travel Adjustments and Lifestyles: A Behavioural Approach. In: Stopher P. & Meyburg A. (Eds.) Behavioral Travel Demand Models, Lexington, Massachusetts, Lexington Books

Reilly, W.J. (1929): Methods for the Study of Retail Relationships, Bulletin 2944. University of Texas, Austin, Texas

Reiner, R., Munz, M. (1990): Ranking Regression Analysis of Spatio-Temporal Variables. Environment and Planning A 22, 507-526

Rijk, F.J.A., Vorst, A.C.F. (1983): On the Existence and Uniqueness of Equilibrium Points in an Urban Retail Model. Environment and Planning A 15, 475-482

Rogers, A. (1966): The Multiregional Matrix Growth Operator and the Stable Interregional Age Structure. Demography 5, 439-442

Rogers, A. (1989): Requiem for the Net Migrant. Geographical Analysis 22, 147-164

Rogerson, P.A., Plane, D.A. (1984): Modelling Temporal Change in Flow Matrices. Papers of the Regional Science Association 54, 147-164

Romer, P.M. (1986): Increasing Returns and Long-Run Growth. Journal of Political Economy 94, 1002-1037

Roy, J.R., Lesse, P.F. (1981): On Appropriate Microstate Descriptions in Entropy Modelling. Transportation Research B 15, 85-96

Roy, J.R. (1983): Estimation of Singly-Constrained Spatial-Interaction Models. Environment and Planning B 10, 269-274

Roy, J.R. (1985): On Forecasting Choice among Dependent Spatial Alternatives. Environment and Planning B 12, 479-492

Roy, J.R., Lesse, P.F. (1985): Entropy Models with Entropy Constraints on Aggregated Events. Environment and Planning A 17, 1669-1674

Roy, J.R., Batten, D.F. (1985): Invited Comment on 'Entropy and Information Theory: Are We Missing Something?'. Environment and Planning A 17, 704-706

Roy, J.R. (1987): An Alternative Information Theory Approach for Modelling Spatial Interaction. Environment and Planning A 19, 385-394

Roy, J.R. (1989): Nested Dynamics and the 4-Step Transport Model. Proceedings of the Fifth World Conference on Transport Research, Yokohama

Roy, J.R. (1990): Unemployment and Unfilled Vacancies in Spatial Models of Job Choice. In: Fischer, M.M., Nijkamp, P., Papageorgiou, Y.Y. (Eds.): Spatial Choices and Processes. North-Holland, Amsterdam New York Oxford Tokyo, 293-312

Roy, J.R., Flood, J. (1992): Interregional Migration Modelling via Entropy and Information Theory. Geographical Analysis 24, 16-34

Roy, J.R. (1993): Comment on "The Development of a New Gravity-Opportunity Model for Trip Distribution". Environment and Planning A 25, 1689-1691

Roy, J.R., Johansson, B. (1993): A Model of Trade Flows in Differentiated Goods. Annals of Regional Science 27, 95-115

Roy, J.R. (1994): Trade With and Without Intermediaries : Some Alternative Model Formulations. Annals of Regional Science 28, 329-343

Roy, J.R. (1995): Dispersed Spatial Input Demand Functions. Annals of Regional Science 29, 375-388

Roy, J.R., Kobayashi, K. (1996): Elements of a Location Model for Knowledge Firms. Invited Paper for the 5th World Congress of the Regional Science Association International, Tokyo (Copy Available from First Author)
Roy, J.R. (1997a): A More Consistent Retail Demand Model: Integration of Discrete and Continuous Choice. Geographical and Environmental Modelling 1, 83-96
Roy, J.R. (1997b): A Price-Responsive Framework for Interregional Input-Output. Annals of Regional Science 31, 285-298
Roy, J.R., Marquez, L.O., Taylor, M.A.P., Ueda, T. (1998): Development of a Compact Urban Simulation Model. In: Lundqvist, L., Mattsson, L.-G., Kim, T.J. (Eds.): Network Infrastructure and the Urban Environment. Springer, Heidelberg, 344-360
Roy, J.R. (1999): Areas, Nodes and Networks. Papers in Regional Science 78, 135-155
Roy, J. R., Smith, N. C., Xu, B. (2001): Simultaneous Modelling of Multi-Purpose/ Multi-Stop Activity Patterns and Quantities Consumed. Geographical Systems 3, 303-324
Samuelson, P.A. (1952a): Spatial Price Equilibrium and Linear Programming. American Economic Review 42, 283-303
Samuelson, P.A. (1952b): The Transfer Problem and Transport Costs. Economic Journal 62, 278-304
Samuelson, P.A. (1983): Thünen at Two Hundred. Journal of Economic Literature 21, 1468-1488
Schneider, M. (1959): Gravity Models and Trip Distribution Theory. Papers and Proceedings of the Regional Science Association 5, 51-56
Sen, A., Smith, T.E. (1995): Gravity Models of Spatial Interaction Behaviour. Springer, Berlin Heidelberg New York
Shannon, C.E. (1948): A Mathematical Theory of Communication. Bell System Technical Journal 27, 379-423, 623-656
Sharpe, R., Roy, J.R. (1985): Teleshopping and Retail Location. In: Brotchie, J.F., Newton, P.W., Hall, P., Nijkamp, P. (Eds.): The Future of Urban Form. Croom Helm, London, Sydney, 188-205
Smith, T.E. (1978): A Cost-Efficiency Principle of Spatial Interaction Behaviour. Regional Science and Urban Economics 8, 313-337
Smith, T.E. (1989): A Simple Decision Theory of Spatial Interaction: The Alonso Model Re-Visited. Working Paper 135, Department of Regional Science, University of Pennsylvania
Smith, T.E. (1990): Most-Probable-State Analysis: A Method for Testing Probabilistic Theories of Population Behaviour. In: Chatterji, M., Kuenne, R.E. (Eds.): New Frontiers in Regional Science, MacMillan, 75-94
Snickars, F., Weibull, J.W. (1977): A Minimum Information Principle: Theory and Practice. Regional Science and Urban Economics 7, 137-168
Southworth, F. (1985): On Household Travel Circuit Benefits and Their Locational Implications. In: Hutchinson, B.G., Nijkamp, P., Batty, M. (Eds.): Optimization and Discrete Choice in Urban Systems. Springer, Heidelberg, 116-131
Stewart, J.Q. (1941): An Inverse Distance Variation for Certain Social Distances. Science 93, 89-90
Stouffer, S.A. (1940): Intervening Opportunities: A Theory Relating Mobility and Distance. American Sociological Review 5, 845-867
Tabuchi, T. (1986): Urban Agglomeration Economies in a Linear City. Regional Science and Urban Economics 16, 421-436
Takayama, T., Judge, G.G. (1964): Equilibrium among Spatially Separated Markets. Econometrica 32, 510-524
Takayama, T., Judge, G.G. (1971): Spatial and Temporal Price and Allocation Models. North-Holland, Amsterdam

Theil, H. (1967): Economics and Information Theory, North Holland, Amsterdam
Thill, J.C. (1995): Modeling Store Choices with Cross-Sectional and Pooled Cross-Sectional Data: A Comparison. Environment and Planning A 27, 1303-1315
Thill, J.C., Horowitz, J.L. (1997): Modeling Non-Work Destination Choices with Choice Sets Defined by Travel-Time Constraints. In Getis, A., Fischer, M.M. (Eds.): Recent Developments in Spatial Analysis - Spatial Statistics, Behavioural Modelling and Neurocomputing, Springer, Heidelberg, 186-208
Thill, J.C., Mozolin, M. (2000): Feed-Forward Neural Networks for Spatial Interaction: Are They Trustworthy Forecasting Tools?. In Reggiani, A. (Ed.): Spatial Economic Science: New Frontiers in Theory and Methodology, Springer, Heidelberg, 355-381
Thünen von, J.H. (1826): Der Isolierte Staat in Beziehung auf Landwirtschaft und Nationalökonomie. Stuttgart (English Translation: (1966): The Isolated State. Pergamon Press, Oxford
Ueda, T., Nakamura, H., Shimizu, E. (1996) GIS Integrated Systems for Urban Transport Development Planning. In: Hayashi, Y., Roy, J.R. (Eds.): Transport, Land-Use and the Environment, Kluwer Academic Publishers, Dordrecht
Van Dijk, J., Oosterhaven, J. (1986): Regional Impacts of Migrants' Expenditure: An Input-Output/Vacancy-Chain Approach. In: Batey, P.W.J., Madden, M. (Eds.): Integrated Analysis of Regional Systems, 122-147
Wakabayashi, M., Hewings, G.J.D. (2002): Life Cycle Changes in Consumption Behavior: Age-Specific and Regional Variations. Discussion Paper 02-T-13, Regional Economics Applications Laboratory, University of Illinois, Urbana
Weidlich, W., Haag, G. (1988): The Migratory Equations of Motion. In: Weidlich, W., Haag, G. (Eds.): Interregional Migration, Springer Verlag, Berlin
Wilson, A.G. (1967): A Statistical Theory of Spatial Distribution Models. Transportation Research 1, 253-269
Wilson, A.G. (1970): Entropy in Urban and Regional Modelling. Pion, London
Young, E.C. (1924): The Movement of Farm Population, Bulletin 426. Cornell Agricultural Experimental Station, Ithaca, New York

Subject Index

Titles in the Series (continued)

M. M. Fischer and *J. Fröhlich* (Eds.)
Knowledge, Complexity and Innovation Systems
XII, 477 pages. 2001. ISBN 3-540-41969-1

M. M. Fischer, J. Revilla Diez and *F. Snickars*
Metropolitan Innovation Systems
VIII, 270 pages. 2001. ISBN 3-540-41967-5

L. Lundqvist and *L.-G. Mattsson* (Eds.)
National Transport Models
VIII, 202 pages. 2002. ISBN 3-540-42426-1

J. R. Cuadrado-Roura and *M. Parellada* (Eds.)
Regional Convergence in the European Union
VIII, 368 pages. 2002. ISBN 3-540-43242-6

G. J. D. Hewings, M. Sonis and *D. Boyce* (Eds.)
Trade, Networks and Hierarchies
XI, 467 pages. 2002. ISBN 3-540-43087-3

G. Atalik and *M. M. Fischer* (Eds.)
Regional Development Reconsidered
X, 220 pages. 2002. ISBN 3-540-43610-3

Z. J. Acs, H. L. F. de Groot and *P. Nijkamp* (Eds.)
The Emergence of the Knowledge Economy
VII, 388 pages. 2002. ISBN 3-540-43722-3

R. J. Stimson, R. R. Stough and *B. H. Roberts*
Regional Economic Development
X, 397 pages. 2002. ISBN 3-540-43731-2

S. Geertman and *J. Stillwell* (Eds.)
Planning Support Systems in Practice
XII, 578 pages. 2003. ISBN 3-540-43719-3

B. Fingleton (Ed.)
European Regional Growth
VIII, 435 pages. 2003. ISBN 3-540-00366-5

T. Puu
Mathematical Location and Land Use Theory,
2nd Edition
X, 362 pages. 2003. ISBN 3-540-00931-0

J. Bröcker, D. Dohse and R. Soltwedel (Eds.)
Innovation Clusters and Interregional Competition
VIII, 409 pages. 2003. ISBN 3-540-00999-X

D. A. Griffith
Spatial Autocorrelation and Spatial Filtering
XIV, 247 pages. 2003. ISBN 3-540-00932-9

Druck: betz-druck GmbH, D-64291 Darmstadt
Verarbeitung: Buchbinderei Schäffer, D-67269 Grünstadt

GPSR Compliance

The European Union's (EU) General Product Safety Regulation (GPSR) is a set of rules that requires consumer products to be safe and our obligations to ensure this.

If you have any concerns about our products, you can contact us on ProductSafety@springernature.com

In case Publisher is established outside the EU, the EU authorized representative is:

Springer Nature Customer Service Center GmbH
Europaplatz 3
69115 Heidelberg, Germany

Batch number: 10704039

Printed by Printforce, the Netherlands